Loop Quantum Gravity for the Bewildered

Sundance Bilson-Thompson

Loop Quantum Gravity for the Bewildered

The Self-Dual Approach Revisited

Second Edition

 Springer

Sundance Bilson-Thompson
School of Physics, Chemistry and Earth
Sciences
University of Adelaide
Adelaide, SA, Australia

With Contribution by
Deepak Vaid
Department of Physics
National Institute of Technology
Mangalore, Karnataka, India

ISBN 978-3-031-43451-8   ISBN 978-3-031-43452-5   (eBook)
https://doi.org/10.1007/978-3-031-43452-5

1st edition: © Springer International Publishing Switzerland 2017
2nd edition: © Springer Nature Switzerland AG 2024

This Springer imprint is published by the registered company Springer Nature Switzerland AG
The registered company address is: Gewerbestrasse 11, 6330 Cham, Switzerland

Paper in this product is recyclable.

Preface

The first edition of this book arose out of a desire to better understand the foundations of loop quantum gravity, at a time when I and my co-author Deepak Vaid were both postdocs. We were essentially exploring a realm of knowledge that was comparatively new to both of us. Although the broad outline of this realm was evident, a cohesive picture of its structure appeared to be hidden behind a substantial mathematical paraphernalia. It wasn't always clear how much of this was necessary to grasp the physical underpinnings of the theory. Towards the goal of improving our own understanding we conceived of and composed a review article in the hopes that by clearing our own bewilderment about topics such as holonomies and spin networks, we might also contribute to mapping out these concepts and the relationships between them for others interested in this rapidly developing field. It was a happy surprise when we were invited to turn that review article into a book, which became the first edition of this text. As our efforts were geared towards understanding the basic structure of loop quantum gravity (LQG) and connecting it to more familiar branches of modern physics, there were many important topics that we left out of the first edition. Perhaps most significantly, we chose to omit any meaningful discussion of the covariant $3+1$ formalism of spacetime dynamics which goes by the name of spin foams.

It was therefore another happy surprise when the first edition was sufficiently well-received that our publishers invited us to produce an updated second edition, affording the opportunity to include a more thorough discussion of the topics we had skipped the first time around. Unfortunately owing to other concerns and personal priorities, Deepak concluded that he was unable to contribute the time and effort needed to bring the second edition to completion. We discussed what role (if any) he wanted to have, and eventually he made the difficult decision to withdraw from co-authorship of the second edition. I wish to make it clear, however, that the work he put into the first edition was inestimable, and without it the current edition would not exist. Furthermore the content of the updates included in this edition was largely inspired by his enthusiasm for these topics, and much of the discussion of group field theory was based upon rough notes that he shared with me. He also kindly volunteered to proof-read the draft of the second edition. Those sections which have not been substantially updated from the first edition still refer to the authors, plural, as they were a joint effort. It is important and fair to ensure that Deepak's contributions are fully recognised. Naturally, any errors or flaws in the second edition are my responsibility alone.

And so, after several years and the devotion of much time and effort towards deeper understanding of the topics at hand, it is a pleasure to present the second edition of this work. This book is not intended to be an account of the current state of the art, as research is always developing but books are static and rapidly fall out-of-date. So while extra material has been included in this edition, the focus has been on broadening the range of foundational concepts covered and emphasising the conceptual links between topics. It is my sincere hope that much as the first edition was successful (at least in its creators' opinion) in providing a pedagogical introduction to the foundations of LQG, the second edition will help clarify the central concepts and technical tools of spin foams, group field theory, spinorial LQG, and coherent states for the reader bewildered and discouraged by the vast literature on these topics and the accompanying mathematical tools which appear to be necessary for these constructions, as well as covering some earlier developments to show where the state of the art comes from.

The opportunity has also been taken to make some minor but desirable amendments, to reflect the intended role of this book as an introduction for interested outsiders from a wide range of backgrounds. These have included a more thorough discussion of the connection between entropy and information theory, a more detailed discussion of the group theory describing special relativity, a more extensive discussion of differential forms to bridge the gap between indexed and index-free notation, as well as a number of adjustments to wording in several instances, additional discussions to enhance the reader's intuition for the meaning behind the maths, and updated illustrations more suitable for viewing in e-book format. Furthermore, the choice of notation has been reworked to ensure the use of symbols and indices is more consistent throughout the text. There is of course much more that could have been included, but I hope the right balance has been struck between the breadth and depth of the content.

Whether or not this book is successful in the goal Deepak and I initially set ourselves for this edition will be determined by its readers. If ever the possibility of a third edition is raised, I will take that as a sign that this work did indeed help some in the community enhance their understanding of this important approach towards a theory of quantum gravity.

I wish again to thank Deepak Vaid for his work on the first edition, as well as his contribution of ideas and feedback on drafts of this edition. I also wish to thank Christian Caron, Chandra Sekaran Arjunan, Lisa Scalone, and Sivananth Sivachandran at Springer for their substantial patience and guidance during the preparation of the manuscript. During the writing of this book, I have been employed in teaching and research positions at the University of Adelaide and RMIT University and was supported in part by the Australian Research Council through a Discovery Project (project number DP200102152). I wish to thank Nicolas Menicucci for his support, discussions, and suggestions, and Valentina Baccetti for her support through the RMIT Vice-Chancellor's Research Fellowship. I also wish to thank Gavin Brennen of Macquarie University for discussions and suggestions, Sebastian

Murk of the Okinawa Institute of Science and Technology for insightful feedback and careful proofreading, and Anthony Williams of the University of Adelaide for his ongoing support.

My parents, Lianne and Gary, have provided several decades' worth of support for my passion for physics, as well as much-needed enthusiasm and encouragement throughout the writing process, and for these and so many other reasons they have my gratitude and love.

Adelaide, Australia   Sundance Bilson-Thompson
September 2023

Conventions

Before we proceed, a quick description of our conventions will hopefully be useful to the reader. We have attempted to strike a balance between clarity of description and simplicity of notation. In particular, the use of different symbols for indices should clearly indicate the number of dimensions, and whether we are dealing with tensors, matrices, Lie algebra generators, etc. Deviations from these conventions may occur from time to time, in the effort to avoid confusion through the re-use of similar symbols. However, every effort has been made to avoid assigning different meanings to the same symbol within individual chapters;

Indices

- Greek letters $\mu, \nu, \rho, \ldots \in \{0, 1, 2, 3\}$ from the middle of the alphabet are four-dimensional spacetime indices. Other Greek letters, $\alpha, \beta, \ldots$ will be used for general cases in N dimensions, and when unused indices are needed to complete tensor expressions.
- Lowercase letters $a, b, c, \ldots$ from the start of the Latin alphabet are integers, primarily used as three-dimensional spatial indices, taking values in $\{1, 2, 3\}$. These will often be used when dealing exclusively with the spatial part of a four-dimensional quantity that would otherwise have Greek indices. More generally, they may be used as integer parameters.
- Lowercase letters $i, j, k, \ldots \in 1, 2, 3, \ldots, N$ from the middle of the Latin alphabet are indices for a space of N dimensions. Equations involving these indices are the general cases, which can be applied to Minkowski space, $\mathbb{R}^3$, etc. They will also sometimes be used as indices labelling rows and columns of matrices, and as $\mathfrak{su}(2)$ Lie algebra indices.
- Uppercase letters $I, J, K, \ldots$ are specifically "internal" indices used for Lie algebra elements, which take values in the appropriate range, such as $\{1, 2, 3\}$ for the Pauli matrices, or $\{0, 1, 2, 3\}$ for the $\mathfrak{sl}(2, \mathbb{C})$ Lorentz Lie algebra.

Symbols

Wherever possible, to avoid ambiguity we will refrain from using "special" letters as indices, or distinguish particular symbols from their index versions to avoid confusion, as follows;

- The Greek lower-case λ refers to paths, and is not used as an index.
- The imaginary unit will be in boldface, hence $\mathbf{i} = \sqrt{-1}$, to distinguish it from the index i.
- Similarly, the bold $\boldsymbol{j}$ will be used to denote spin, when necessary, to distinguish it from an index.
- The cosmological scale factor will be denoted in bold, hence $\boldsymbol{a}$, to distinguish it from an index. We follow the standard practice of using a for lattice spacing, and a, b, c for the sides of a triangle; however, it should be clear from context when these are not indices.
- The Greek lower-case γ without indices occurs as an index in the early chapters, but from Chap. 5 onwards it is used exclusively for the Barbero-Immirzi parameter. With indices (e.g. γ^μ or γ_I), it always denotes the Dirac gamma matrices.
- Annihilation and creation operators are denoted by $\hat{a}$, $\hat{b}$ (and their conjugates), rather than a, b.
- For blades and forms (see Appendix B), the letters u, v, w represent k-blades, and an arrow above a symbol (e.g. $\vec{u}$, $\vec{V}$) denotes a vector, in any number of dimensions. Differential forms will be represented either by α, β or Roman capitals as noted in the text.
- The gravitational constant is denoted by the symbol $\mathcal{G}$, rather than G.

Contents

1 Introduction 1
  1.1 Motivation and Some History 1
  1.2 Overview: Loop Quantum Gravity and Friends 4
  References 6

2 Classical GR 9
  2.1 Parallel Transport and Curvature 10
  2.2 Einstein's Field Equations 13
  2.3 Coordinates and Diffeomorphism Invariance 14
  References 17

3 Quantum Field Theory 19
  3.1 Covariant Derivative and Curvature 19
  3.2 Dual Tensors, Bivectors and k-forms 23
  3.3 Wilson Loops and Holonomies 26
  3.4 Dynamics of Quantum Fields 29
  References 32

4 Expanding on Classical GR 33
  4.1 Lagrangian Approach: The Einstein-Hilbert Action 33
  4.2 Hamiltonian Approach: The ADM Splitting 36
     4.2.1 Physical Interpretation of Constraints 43
  4.3 Seeking a Path to Canonical Quantum Gravity 44
     4.3.1 Connection Formulation 45
     4.3.2 Tetrads 47
     4.3.3 Choosing a Gauge Group 48
     4.3.4 Spin Connection 51
     4.3.5 Palatini Action 53
     4.3.6 Palatini Hamiltonian and Constraints 55
  References 57

5 First Steps to a Theory of Quantum Gravity 59
  5.1 Ashtekar Formulation: "New Variables" for General
     Relativity 60
  5.2 The Barbero-Immirzi Parameter 62
  5.3 To Be or Not to Be (Real) 63

5.4   Loop Quantization 65
5.5   Canonical Quantization 65
References 66

6 Kinematical Hilbert Space 69
6.1   Kinematics via a Toy Model 69
6.2   Space of Generalised Connections 77
6.3   Area Operator 79
6.4   Volume Operator 84
6.5   Spin Networks 87
6.6   Looking Ahead to Spin Foams 88
References 88

7 Dynamics of Spin Networks 91
7.1   Spin Foams 92
        7.1.1   Transition Amplitudes 96
7.2   Early Developments 97
        7.2.1   BF Theory 97
        7.2.2   Chern–Simons Theory 99
        7.2.3   The Cosmological Constant 101
7.3   Some Recent Developments 104
        7.3.1   Tensor Networks 104
        7.3.2   Spinorial LQG and Coherent States 106
        7.3.3   Group Field Theory 111
References 116

8 Applications 119
8.1   Black Hole Entropy 120
        8.1.1   Rovelli's Counting 124
        8.1.2   Number Theoretical Approach 125
        8.1.3   Chern-Simons Approach 127
        8.1.4   Entropy from Entanglement 127
8.2   Loop Quantum Cosmology 129
        8.2.1   Isotropy and Homogeneity in the Metric Formulation 130
        8.2.2   FLRW Models 130
        8.2.3   Connection Variables 132
        8.2.4   Holonomy Variables 133
        8.2.5   Quantisation 134
        8.2.6   Triad Eigenstates and Volume Quantization 135
        8.2.7   Regularized FLRW Hamiltonian 136
        8.2.8   Singularity Resolution and Bouncing Cosmologies 137
References 139

9 Discussion 143
References 145

Appendix A: Groups, Representations, and Algebras 147

Appendix B: Blades, Forms, and Duality 153

Appendix C: Path Ordered Exponential 167

Appendix D: ADM Variables 169

Appendix E: Lie Derivative 173

Appendix F: 3+1 Decomposition of the Palatini Action 175

Appendix G: The Kodama State 179

Appendix H: Peter-Weyl Theorem 181

Appendix I: Regge Calculus 183

Appendix J: Fibre Bundles 185

Appendix K: Knots, Links, and the Kauffman Bracket 189

Appendix L: Quantum (or q-Deformed) Groups 195

Appendix M: Entropy 199

Appendix N: Square-Free Numbers 205

Appendix O: Brahmagupta-Pell Equation 207

Acronyms

AdS/CFT	Anti-de Sitter/Conformal Field Theory
AL	Ashtekar-Lewandowski
BH	Bekenstein-Hawking
BP	Brahmagupta-Pell
CDT	Causal Dynamical Triangulations
EFEs	Einstein field equations
EHA	Einstein-Hilbert-Ashtekar
FLRW	Friedmann-LeMaitre-Robertson-Walker
GR	General Relativity
GUTs	Grand Unified Theories
KMS	Kubo-Martin-Schwinger
LQC	Loop Quantum Cosmology
LQG	Loop Quantum Gravity
NPP	Number partitioning problem
QCD	Quantum Chromodynamics
QED	Quantum Electrodynamics
QFT	Quantum Field Theory
QIH	Quantum isolated horizon
RS	Rovelli-Smolin
SM	Standard Model
SR	Special Relativity

Introduction

1

1.1   Motivation and Some History

The goal of Loop Quantum Gravity (LQG) is to take two extremely well-developed and successful theories, General Relativity (GR) and Quantum Field Theory (QFT), at "face value" and attempt to combine them into a single theory with a minimum of assumptions and deviations from established physics. Our goal, as authors of this book, is to provide a succinct but clear description of LQG—the main body of concepts in the current formulation of LQG relying primarily on the self-dual variables approach, some of the historical basis underlying these concepts, and a few simple yet interesting results—aimed at the reader who has more curiosity than familiarity with the underlying concepts, and hence desires a broad, pedagogical overview before attempting to read more technical discussions. This book is inspired by the view that one never truly understands a subject until one tries to explain it to others. Accordingly we have attempted to create a discussion which we would have wanted to read when first encountering LQG. Everyone's learning style is different, and accordingly we make note of several other reviews of this subject [1–11], which the reader may refer to in order to gain a broader understanding, and to sample the various points of view held by researchers in the field.

We will begin with a brief review of the history of the field of quantum gravity in the remainder of this section. Following this we provide a non-technical overview of some of the models and theories that are similar to LQG, in the interests of showing the reader some of the "landscape" of approaches to formulating a viable theory of quantum gravity. We will return to some of these ideas near the end of the text, once the technical details of LQG have been discussed. To begin our discussion of LQG proper, we review some topics in general relativity in Chap. 2 and quantum field theory in Chap. 3, which hopefully fall into the "Goldilocks zone", providing all the necessary basis for LQG, and nothing more. We may occasionally introduce concepts in greater detail than the reader considers necessary, but we feel that when introducing concepts to a (hopefully) wide audience who find them unfamiliar,

© Springer Nature Switzerland AG 2024
S. Bilson-Thompson, *Loop Quantum Gravity for the Bewildered*,
https://doi.org/10.1007/978-3-031-43452-5_1

insufficient detail is more harmful than excessive detail. We will discuss the Lagrangian and Hamiltonian approaches to classical GR in more depth and set the stage for its quantization in Chap. 4 then sketch a conceptual outline of the broad program of quantization of the gravitational field in Chap. 5, before moving on to our main discussion of the loop quantum gravity approach in Chap. 6. The pros and cons of the self-dual variables are also discussed in some depth in Sect. 5.3. In Chap. 8 we cover applications of the ideas and methods of LQG to the counting of microstates of black holes (Sect. 8.1) and to the problem of quantum cosmology (Sect. 8.2). Last, but not least, some common criticisms of LQG and our rebuttals thereof are presented in Chap. 9 along with a discussion of its present status and future prospects, and we discuss again (in more technical detail) some similar theories that may (in the course of future developments) borrow from, lend ideas to, or subsume LQG.

It is assumed that the reader has a minimal familiarity with the tools and concepts of differential geometry, quantum field theory, and general relativity, though we aim to remind the reader of any relevant technical details as necessary.[1] In light of the extensive use of concepts from group theory throughout the text, a brief summary of important concepts and notation relevant to Lie groups and Lie algebras is provided in Appendix A. While this material is widely covered in other texts and undergraduate courses, we feel it's better to have it on hand for reference and not need it, than to need it and not have it.

Before we begin, it would be helpful to give the reader a historical perspective of the developments in theoretical physics which have led us to the present stage.

We are all familiar with classical geometry consisting of points, lines and surfaces. The framework of Euclidean geometry provided the mathematical foundation for Newton's work on inertia and the laws of motion. In the 19th century Gauss, Riemann and Lobachevsky, among others, developed notions of *curved* geometries in which one or more of Euclid's postulates were loosened. The resulting structures allowed Einstein and Hilbert to formulate the theory of general relativity which describes the motion of matter through spacetime as a consequence of the curvature of the background geometry. This curvature in turn is induced by the matter content as encoded in Einstein's equations (2.10). Just as the parallel postulate was the unstated assumption of Newtonian mechanics, whose rejection led to Riemmanian geometry, the unstated assumption underlying the framework of general relativity is that of the smoothness and continuity of spacetime on all scales.

Loop quantum gravity and related approaches invite us to consider that our notion of spacetime as a smooth continuum must give way to an atomistic description of geometry in which the classical spacetime we observe around us emerges from the interactions of countless (truly indivisible) *atoms* of spacetime. This idea is grounded in mathematically rigorous results, but is also a natural continuation of the trend that began when 19th century attempts to reconcile classical thermodynamics

---

[1] Given that we are aiming this book at a broad audience, we may even hope that some readers will find it helpful with their understanding of GR and/or QFT, quite aside from its intended role explaining quantum gravity.

with the physics of radiation encountered fatal difficulties—such as James Jeans' "ultraviolet catastrophe". These difficulties were resolved only when work by Planck, Einstein and others in the early 20th century provided an atomistic description of electromagnetic radiation in terms of particles or "quanta" of light known as *photons*. This development spawned quantum mechanics, and in turn quantum field theory, while around the same time the special and general theories of relativity were being developed.

In the latter part of the 20th century physicists attempted, without much success, to unify the two great frameworks of quantum mechanics and general relativity. For the most part it was assumed that gravity was a phenomenon whose ultimate description was to be found in the form of a quantum field theory as had been so dramatically and successfully accomplished for the electromagnetic, weak and strong forces in the framework known as the Standard Model (SM). These three forces could be understood as arising due to interactions between elementary particles mediated by gauge bosons whose symmetries were encoded in the groups U(1), SU(2) and SU(3) for the electromagnetic, weak and strong forces, respectively. The universal presumption was that the final missing piece of this "grand unified" picture, gravity, would eventually be found as the QFT of some suitable gauge group. This was the motivation for the various Grand Unified Theories (GUTs) developed by Glashow, Pati-Salam, Weinberg and others where the hope was that it would be possible to embed the gravitational interaction along with the Standard Model in some larger group (such SO(5), SO(10) or E_8 depending on the particular scheme). Such schemes could be said to be in conflict with Occam's dictum of simplicity and Einstein and Dirac's notions of beauty and elegance. *More importantly all these models assumed implicitly that spacetime remains continuous at all scales.* As we shall see this assumption lies at the heart of the difficulties encountered in unifying gravity with quantum mechanics.

A significant obstacle to the development of a theory of quantum gravity is the fact that GR is not renormalizable. The gravitational coupling constant $\mathcal{G}$ (or equivalently $1/M_{\text{Planck}}^2$ in dimensionless "natural" units where $\mathcal{G} = c = \hbar = 1$) is not dimensionless, unlike the fine-structure constant α in Quantum Electrodynamics (QED). This means that successive terms in any perturbative series have increasing powers of momenta in the numerator. Rejecting the notion that systems could absorb or transmit energy in arbitrarily small amounts led to the photonic picture of electromagnetic radiation and the discovery of quantum mechanics. Likewise, rejecting the notion that spacetime is arbitrarily smooth at all scales—and replacing it with the idea that geometry at the Planck scale must have a discrete character—leads us to a possible resolution of the ultraviolet infinities encountered in quantum field theory and to a theory of "quantum gravity".

Bekenstein's observation [12–14] of the relationship between the entropy of a black hole and the area of its horizon combined with Hawking's work on black hole thermodynamics led to the realization that there were profound connections between thermodynamics, information theory and black hole physics. These can be succinctly summarized by the famous *area law* relating the Bekenstein–Hawking (BH) entropy of a *macroscopic* black hole S_{BH} to its surface area A:

$$S_{BH} = \frac{k_B A}{4 l_P^2} = \frac{k_B c^3 A}{4 \mathcal{G} \hbar} \qquad (1.1)$$

with l_P^2 being the Planck area and k_B the Boltzmann constant (which is sometimes set to 1). While a more detailed discussion will wait until Sect. 8.1, we note here that if geometrical observables such as area are quantized Eq. (1.1) can be seen as arising from the number of ways that one can join together N quanta of area to form a horizon. In LQG the quantization of geometry arises naturally—though not all theorists are convinced that geometry should be quantized or that LQG is the right way to do so.

1.2   Overview: Loop Quantum Gravity and Friends

As just discussed, LQG is an attempt to describe gravity in a manner that is compatible with both general relativity and quantum field theory. However, it is by no means the only attempt to do so. LQG is often compared, and set in opposition to, string theory. But it is perhaps fairer to characterise the various quantum gravity schemes as attempts to create a gravitational theory of quantum fields (i.e. starting with quantum field theory and constructing a theory that recreates GR in some limit), and attempts to create a quantum theory of gravitational fields (i.e. starting with GR and attempting to quantise it). The former approach includes string theory and related ideas, while the latter approach includes loop quantum gravity and related ideas.

At the time of writing there is no compelling experimental evidence to support a preference for one approach over the other. The various attempts to create a theory of quantum gravity show promise in some regards, and face difficulties in their own ways. It is possible that a successful theory of quantum gravity will incorporate features of several existing approaches. To paraphrase Carlo Rovelli, at this stage humanity is not trying to decide between viable candidate theories of quantum gravity—quite the opposite. Just finding one candidate that is compatible with experimental data would be a significant advance! But at the same time this gives us hope that the space of possible theories is not overwhelming, and the requirement of consistency with quantum mechanics and general relativity should be a reliable guide to our goal.

For the sake of introducing the concepts which will be developed here, it seems appropriate to start with a nontechnical overview of some of the ideas and terminology related to loop quantum gravity.

String theory falls into the first category mentioned above—that of trying to create a gravitational theory of quantum fields. It postulates a fixed, continuous background spacetime,[2] and fundamental entities existing within this background which are

---

[2] Some attempts have been made to develop background-independent formulations, see for instance Smolin's contribution to [15].

extended rather than pointlike. The various topologies and quantised vibrations of these "strings" give rise to a range of fermions and bosons, and the claim that string theory is a theory of quantum gravity rests upon the idea that the bosons which occur amongst these vibrations include the graviton. The quantisation procedure in string theory implies the existence of several dimensions beyond the three spatial and one temporal dimension of everyday experience, explaining this disparity through the idea that the extra dimensions are rolled up or "compactified" to a length scale well below detection with current experiments. The number of extra dimensions necessary is also modified by the incorporation of supersymmetry, leading to the name "superstrings".

Loop quantum gravity falls into the second category—that of creating a quantum theory of gravitational fields. In general the starting point for this approach is the idea that connections and paths between different points in spacetime are more important than the coordinates of the points. Curvature of spacetime corresponds to gravitational fields, and this curvature is both dynamical, and independent of any embedding in a background space. This is what we meant when we said in Sect. 1.1 that GR is taken "at face value". Perhaps the most explicit example of the idea that relations between points in spacetime are more important than coordinates is causal set theory. This starts with a set of points and a partial ordering (i.e. a sense in which some pairs of points, but not all, can be compared to see which point comes "before" the other), and attempts to build the structure of spacetime from these ingredients, developing concepts of geometry, geodesics, and so forth from the combinations of the interrelations between points. Causal sets are similar in many ways to Penrose's conception of spin networks, which were created with the goal of building spacetime structure entirely from combinatorial rules. While it is extremely easy to take a smooth manifold and approximate it by a set of points, defining a smooth manifold from an arbitrarily-generated set of points—sometimes referred to as the "inverse problem"—has presented a considerable challenge within causal set theory for some time (see, for instance, Stachel and Smolin's respective contributions to [15].) Another approach with a similar goal is Causal Dynamical Triangulations (CDT) [16]. These are an outgrowth of Regge calculus (Appendix I), which aimed to simplify calculations in general relativity by approximating spacetime by a collection of small sections, or simplices. Dynamical triangulations developed from this in an attempt to see if an n-dimensional spacetime could develop (subject to certain dynamical rules) from an $(n-1)$-dimensional collection of simplices, but it was discovered that without a sense of causality (such that edges which have an associated time direction are oriented to agree with each other) the resulting structures were unphysical. This may be seen as another manifestation of the inverse problem that occurs in causal set theory. With the imposition of causality, CDT has shown itself to be capable of numerically simulating the formation of manifolds (with a discrete structure) that appear to grow like inflating "mini-universes", and approximate 4-dimensional spacetime at large scales.

Loop quantum gravity and spin foams are often mentioned in the same breath, but it is worth recognising that they developed somewhat in parallel. Loop quantum gravity was infused with the concept of spin networks, creating a viable set of kinematical states to describe quantum spacetime. Spin foams grew out of concepts like BF theory (Sect. 7.2), providing a concept of dynamics in need of a kinematics to complete it—a role which LQG has stepped in to fill. We shall mention almost all of these ideas, and more, in greater detail throughout the rest of this book. But hopefully this overview serves to familiarise the reader with the names of various theories and models, and show that there is a lot to learn, and that many more discoveries lie ahead. For the interested reader, a more thorough survey of these ideas is provided by Smolin in [15]. The history of the quest for a viable theory of quantum gravity, like all of science, is one of ideas converging, combining serendipitously, and spawning new and unexpected avenues for exploration. Even loop quantum gravity and string theory (as just noted, so frequently cast as opponents) show signs of finding common ground in recent years.

With this historical summary and overview in mind, it is now worth covering the basic notions of general relativity and QFT before we attempt to see how these two disciplines may be unified in a single framework.

References

1. M. Gaul, C. Rovelli, Loop quantum gravity and the meaning of diffeomorphism invariance. Lect. Notes. Phys. **541**, 277–324 (2000). https://doi.org/10.48550/arXiv.gr-qc/9910079. arXiv: gr-qc/9910079v2
2. A. Ashtekar, J. Lewandowski, Background independent quantum gravity: a status report. Class. Quant. Grav. **21**(15), R53–R152 (2004). https://doi.org/10.1088/0264-9381/21/15/R01. arXiv: gr-qc/0404018
3. C. Kiefer, Quantum gravity: general introduction and recent developments. Annalen Phys. **15**, 129–148 (2005). https://doi.org/10.1002/andp.200510175. (arXiv: gr-qc/0508120)
4. H. Nicolai, K. Peeters, Loop and spin foam quantum gravity: a brief guide for beginners. Lect. Notes. Phys. **721**, 151–184 (2007). https://doi.org/10.1007/978-3-540-71117-9_9. (arXiv: hep--th/0601129)
5. S. Alexandrov, P. Roche, Critical overview of loops and foams. Phys. Rept. **506**, 41–86 (2011). https://doi.org/10.1016/j.physrep.2011.05.002. (arXiv: 1009.4475)
6. S. Mercuri, Introduction to loop quantum gravity. PoS ISFTG 016 (2009). https://doi.org/10.48550/arXiv.1001.1330. arXiv: 1001.1330
7. P. Doná, S. Speziale, Introductory lectures to loop quantum gravity (2010). https://doi.org/10.48550/arXiv.1007.0402. arXiv: 1007.0402
8. G. Esposito, An introduction to quantum gravity (2011). https://doi.org/10.48550/arXiv.1108.3269. arXiv: 1108.3269
9. C. Rovelli, Zakopane Lectures on loop gravity (2011). https://doi.org/10.48550/arXiv.1102.3660. arXiv: 1102.3660
10. A. Ashtekar, Introduction to loop quantum gravity (2012). https://doi.org/10.1007/978-3-642-33036-0_2. arXiv: 1201.4598
11. A. Perez, The new spin foam models and quantum gravity (2012). https://doi.org/10.48550/arXiv.1205.0911. arXiv:1205.0911
12. J. Bekenstein, Black holes and the second law. Lettere Al Nuovo Cimento (1971–1985) **4**(15), 737–740 (1972). ISSN: 0375-930X. http://dx.doi.org/10.1007/BF02757029
13. J.D. Bekenstein, Black holes and entropy. Phys. Rev. D **7**(8), 2333–2346 (1973). http://dx.doi.org/10.1103/PhysRevD.7.2333
14. J.D. Bekenstein, Extraction of energy and charge from a black hole. Phys. Rev. D **7**, 949–953 (1973). https://doi.org/10.1103/PhysRevD.7.949

15. D. Rickles, S. French, J. Saatsi (eds.), *The Structural Foundations of Quantum Gravity* (Clarendon Press, 2006)
16. R. Loll, J. Ambjorn, J. Jurkiewicz, The universe from scratch. Contemp. Phys. **46**, 103–117 (2006). https://doi.org/10.1080/00107510600603344. (arXiv: hep-th/0509010v3)

Classical GR

2

General Relativity (GR) is an extension of Einstein's theory of Special Relativity (SR), which was required in order to include observers in non-trivial gravitational backgrounds. SR applies in the absence of gravity, and in essence it describes the behavior of vector quantities in a four-dimensional Galilean space, with the Minkowski metric[1]

$$\eta_{\mu\nu} = \text{diag}(-1, +1, +1, +1), \tag{2.1}$$

leading to a 4D line-element

$$ds^2 = -c^2 dt^2 + dx^2 + dy^2 + dz^2. \tag{2.2}$$

The speed of a light signal, measured by any inertial observer, is a constant, denoted c. If we denote the components of a vector in four-dimensional spacetime with Greek indices (e.g. v^μ) the Minkowski metric[2] divides vectors into three categories; *timelike* (those vectors for which $\eta_{\mu\nu} v^\mu v^\nu < 0$), *null* or *light-like* (those vectors for which $\eta_{\mu\nu} v^\mu v^\nu = 0$), and *spacelike* (those vectors for which $\eta_{\mu\nu} v^\mu v^\nu > 0$). Any point, with coordinates (ct, x, y, z), is referred to as an *event*, and the set of all null vectors having their origin at any event define the future light-cone and past light-cone of that event (Fig. 2.1). Events having time-like or null displacement from a given event E_0 (i.e. lying inside or on E_0's lightcones) are causally connected to E_0. Those in/on the past light-cone can influence E_0, those in/on the future lightcone can be influenced by E_0.

---

[1] Of course the choice $\text{diag}(+1, -1, -1, -1)$ is equally valid but we will have occasion later to restrict our attention to the spacial part of the metric, in which case a positive (spatial) line-element is cleaner to work with.

[2] Strictly speaking it is a pseudo-metric, as the distance it measures between two distinct points can be zero.

© Springer Nature Switzerland AG 2024
S. Bilson-Thompson, *Loop Quantum Gravity for the Bewildered*,
https://doi.org/10.1007/978-3-031-43452-5_2

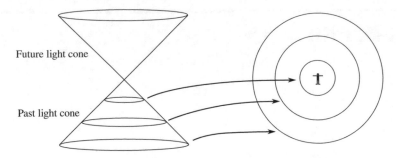

Fig. 2.1 The future-pointing and past-pointing null vectors at an event define the future and past light cones of that event. One spatial dimension is omitted in this diagram, so in fact slices (at constant time) through the past light cone of an observer are two-spheres centred on the observer, and hence map directly to that observer's celestial sphere

General relativity extends these concepts to non-Euclidean spacetime. The metric of this (possibly curved) spacetime is denoted $g_{\mu\nu}$. Around each event (i.e. point in spacetime) it is possible to consider a sufficiently small region that the curvature of spacetime within this region is negligible, and hence the central concepts of special relativity apply locally. Rather than developing the idea that the curvature of space-time gives rise to gravitational effects, we shall treat this as assumed knowledge, and discuss how the curvature of spacetime may be investigated. In general one cannot usefully extend the coordinate system in the region of one event to the region of another arbitrary event since spacetime is not assumed to be flat and Euclidean (we'll define "flat" and "curved" rigorously below). This can be seen from the fact that a Cartesian coordinate system which defined "up" to be the z-axis at one point on the surface of the Earth, would have to define "up" not to be parallel to the z-axis at most other points. In short, a freely-falling reference frame cannot be extended to each point in the vicinity of the surface of the Earth—or any other gravitating body. We are thus forced to work with local coordinate systems which vary from region to region. We shall refer to the basis vectors of these local coordinate systems by the symbols e_j. A set of four such basis vectors at any event is called a *tetrad* or *vierbein*. The discussion of tetrads will be taken up again in Sect. 4.3.2. The metric is related to the dot product of basis vectors by $g_{jk} = e_j \cdot e_k$. In an orthonormal basis with Euclidean metric the convenient relation $e_j \cdot e_k = \delta_{jk}$ holds true. As the basis vectors are not necessarily orthonormal in general, we may define another set of vectors e^j, which satisfy the analogous relationship $e^j \cdot e_k = \delta_k^j$. The e^j will be referred to as the dual basis vectors or covectors.

2.1 Parallel Transport and Curvature

Given the basis vectors e_j of a local coordinate system, an arbitrary vector is written in terms of its components v^j as $\vec{V} = v^j e_j$. It is of course also possible to define vectors with respect to the dual basis. These dual vectors will have components with lowered indices, for example v_j, and take the general form $v_j e^j$. The metric is used

to switch between components referred to the basis or dual basis, e.g. $v_k = g_{jk}v^j$.
Vectors defined with raised indices on their components are called 'contravariant
vectors' or simply 'vectors'. Those with lowered indices are called 'covariant vectors,
'covectors' or '1-forms'.[3] Note that the e_j, having lowered indices, are basis vectors
while the e^j, having raised indices, are basis 1-forms. We will return to the distinction
between vectors and 1-forms in Sect. 3.2.

When we differentiate a vector along a curve parametrised by the coordinate x^k
we must apply the product rule, as the vector itself can change direction and length,
and the local basis will in general also change along the curve, hence

$$\frac{\mathrm{d}\vec{V}}{\mathrm{d}x^k} = \frac{\partial v^j}{\partial x^k}e_j + v^j\frac{\partial e_j}{\partial x^k}. \tag{2.3}$$

We extract the mth component by taking the dot product with the dual basis vector
(i.e. basis 1-form) e^m, since $e^m \cdot e_j = \delta_j^m$. Hence we obtain

$$\frac{\mathrm{d}v^m}{\mathrm{d}x^k} = \frac{\partial v^m}{\partial x^k} + v^j\frac{\partial e_j}{\partial x^k}\cdot e^m\,, \tag{2.4}$$

which by a suitable choice of *notation* is usually rewritten in the form

$$\nabla_k v^m = \partial_k v^m + v^j\Gamma^m{}_{jk}. \tag{2.5}$$

The derivative written on the left-hand-side is termed the *covariant derivative*, and
consists of a partial derivative due to changes in the vector, and a term $\Gamma^m{}_{jk}$ called
the *connection* which is due to changes in the local coordinate basis from one place to
another. The symbols $\Gamma^m{}_{jk}$ are called Christoffel symbols, and so in later chapters we
will frequently refer to this term as the Christoffel connection. If a vector is parallel-
transported along a path, its covariant derivative will be zero. In consequence any
change in the components of the vector is due to (and hence equal and opposite to)
the change in local basis, so that

$$\frac{\partial v^m}{\partial x^k} = -v^j\frac{\partial e_j}{\partial x^k}\cdot e^m\,. \tag{2.6}$$

The transport of a vector along a single path between two distinct points does not
reveal any curvature of the space (or spacetime) through which the vector is carried.
To detect curvature it is necessary to carry a vector all the way around a closed path
and back to its starting point, and compare its initial and final orientations. If they

---

[3] The terms covariant and contravariant come about in reference to whether the quantity transforms
the same way as or oppositely to the basis vectors, but applying it to a vector like $\vec{V} = v^j e_j$, which
has no uncontracted indices, is something of a misnomer. It is really the *components*—referred to a
particular choice of basis—that vary, so an expression like 'contravariant (covariant) vector' should
be understood as shorthand for "vector with components defined with respect to a particular basis
(dual basis)."

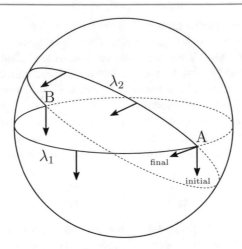

Fig. 2.2 The parallel transport of a vector around a closed path tells us about the curvature of a region bounded by that path. Here a vector is parallel transported along curve λ_1 from A to B, and back from B to A along λ_2. Both λ_1 and λ_2 are sections of great circles, and so we can see that the vector maintains a constant angle to the tangent to the curve between A and B, but this angle changes abruptly at B when the vector switches from λ_1 to λ_2. The difference in initial and final orientation of the vector at A tells us that the surface (a sphere in this case) is curved. Just as an arbitrarily curved path in $\mathbb{R}^2$ can be built up from straight line segments, an arbitrary path in a curved manifold can be built up from sections of geodesics (of which great circles are an example)

are the same, for an arbitrary path, the space (or spacetime) is *flat*. If they differ, the space is *curved*, and the amount by which the initial and final orientations of the vector differ provides a measure of how much curvature is enclosed within the path. Alternatively, one may transport two copies of a vector from the same starting point, A, along different paths, λ_1 and λ_2 to a common end-point, B. Comparing the orientations of the vectors after they have been transported along these two different paths reveals whether the space is flat or curved. It should be obvious that this is equivalent to following a closed path (moving along λ_1 from A to B, and then along λ_2 from B to A, c.f. Fig. 2.2). The measure of how much this closed path (loop) differs from a loop in flat space (that is, how much the two transported vectors at B differ from each other) is called the *holonomy* of the loop. The reader is cautioned that the term 'holonomy' is sometimes used with different meanings in different contexts. To (hopefully) minimise confusion we shall employ the term "loop holonomy" for the concept we have just introduced.

In light of the preceding discussion, suppose a vector $\vec{V}$ is transported from point A some distance in the μ-direction. The effect of this transport upon the components of $\vec{V}$ is given by the covariant derivative ∇_μ of $\vec{V}$. The vector is then transported in the ν-direction to arrive at point B. An identical copy of the vector is carried first from A in the ν-direction, and then in the μ-direction to B. The difference between the two resulting (transported) vectors, when they arrive at B is given by

$$(\nabla_\mu \nabla_\nu - \nabla_\nu \nabla_\mu)\vec{V}. \tag{2.7}$$

This commutator defines the Riemann curvature tensor,

$$R^{\delta}{}_{\rho\mu\nu}v^{\rho} = [\nabla_{\mu}, \nabla_{\nu}]v^{\delta}. \tag{2.8}$$

If and only if the space is flat, all the components of $R^{\delta}{}_{\rho\mu\nu}$ will be zero, otherwise the space is curved.

Since the terms in the commutator of covariant derivatives differ only in the ordering of the indices, it is common to place the commutator brackets around the indices only, rather than the operators, hence we can write

$$\nabla_{[\mu}\nabla_{\nu]} = [\nabla_{\mu}, \nabla_{\nu}] = \nabla_{\mu}\nabla_{\nu} - \nabla_{\nu}\nabla_{\mu}. \tag{2.9}$$

2.2 Einstein's Field Equations

Einstein's equations relate the curvature of spacetime with the energy density of the matter and fields present in the spacetime. Defining the Ricci tensor $R_{\rho\nu} = R^{\mu}{}_{\rho\mu\nu}$ and the Ricci scalar $R = R^{\nu}{}_{\nu}$ (i.e. it is the trace of the Ricci tensor, taken after raising an index using the metric $g^{\mu\nu}$), the relationship between energy density and spacetime curvature is then given by

$$R^{\mu\nu} - \frac{1}{2}Rg^{\mu\nu} + \Lambda g^{\mu\nu} = 8\pi\mathcal{G}T^{\mu\nu}, \tag{2.10}$$

where $\mathcal{G}$ is Newton's constant, and the coefficient Λ is the cosmological constant, which prior to the 1990s was believed to be identically zero. The tensor $T^{\mu\nu}$ is the energy-momentum tensor (also referred to as the stress-energy tensor). We will not discuss it in great detail, but its components describe the flux of energy and momentum (i.e. 4-momentum) across various timelike and spacelike surfaces.[4] The component $T^{\mu\nu}$ describes the flux of the μth component of 4-momentum across a surface of constant x^{ν}. For instance, the zeroeth component of 4-momentum is energy, and hence T^{00} is the amount of energy crossing a surface of constant $time$ (i.e. energy per unit volume that is moving into the future but stationary in space, hence it is the energy density).

It should be noted that we can write $\Gamma^{\rho}{}_{\mu\nu}$ in terms of the metric $g_{\mu\nu}$ (see e.g. [1]),

$$\Gamma^{\rho}{}_{\mu\nu} = \frac{1}{2}g^{\rho\delta}\left(\partial_{\mu}g_{\delta\nu} + \partial_{\nu}g_{\delta\mu} - \partial_{\delta}g_{\mu\nu}\right). \tag{2.11}$$

Since the Riemann tensor is defined from the covariant derivative, and the covariant derivative is defined by the connection, the metric $g^{\mu\nu}$ should be interpreted as a solution of the Einstein field equations (EFEs), Eq. (2.10).

---

[4] The presence of the energy-momentum tensor is related to the fact that it is not merely the mass of matter that creates gravity, but its momentum, as required to maintain consistency when transforming between various Lorentz-boosted frames.

It is sometimes preferable to write Eq. (2.10) in the form

$$G^{\mu\nu} = 8\pi \mathcal{G} T^{\mu\nu} - \Lambda g^{\mu\nu} \tag{2.12}$$

where the *Einstein tensor* $G^{\mu\nu} = R^{\mu\nu} - Rg^{\mu\nu}/2$ is the divergence-free part of the Ricci tensor. The explicit form of Eq. (2.10) emphasises the relationship between mass-energy and spacetime curvature. All the quantities related to the structure of the spacetime (i.e. $R^{\mu\nu}$, R, $g^{\mu\nu}$) are on the left-hand side. The quantity related to the presence of matter and energy, $T^{\mu\nu}$, is on the right-hand side. For now it remains a question of interpretation whether this means that mass-energy is equivalent to spacetime curvature, or identical to it. Perhaps more importantly the form of the EFEs makes it clear that GR is a theory of dynamical spacetime. As matter and energy move, so the curvature of the spacetime in their vicinity changes.

It is worth noting (without proof, see for instance [1]) that the gravitational field in the simplest case of a static, spherically-symmetric field around a mass M, defines a line element of the form derived by Schwarzschild,

$$ds^2 = -c^2 \left(1 - \frac{2\mathcal{G}M}{c^2 r}\right) dt^2 + \left(1 - \frac{2\mathcal{G}M}{c^2 r}\right)^{-1} dr^2 + r^2 (d\theta^2 + \sin^2 \theta d\phi^2). \tag{2.13}$$

For weak gravitational fields, and test masses moving at low velocities (that is, $v \ll c$) the majority of the deviation from the line element in empty space is caused by the coeffcient of the dt^2 term on the right. This situation also coincides with the limit in which Newtonian gravity becomes a good description of the mechanics. In the Newtonian picture the force of gravity can be written as the gradient of a potential,

$$\vec{F} = \nabla V. \tag{2.14}$$

It can be shown that

$$\partial g_{00} \propto \nabla V, \tag{2.15}$$

implying that gravity in the Newtonian or weak-field limit can be understood, primarily, as the amount of distortion in the local "speed" of time caused by the presence of matter—an effect known as gravitational time dilation.

2.3   Coordinates and Diffeomorphism Invariance

General relativity embodies a principle called *diffeomorphism invariance*. This principle states, in essence, that the laws of physics should be invariant under different choices of coordinates. In fact, one may say that coordinates have no meaning in the formulation of physical laws, and in principle we could do without them.

In a practical sense, however, when performing calculations it is often necessary to work with a particular choice of coordinates. When translating in spacetime we may find that the basis vectors are defined differently at different points (giving rise to a connection, as we saw above). However if we restrict our attention to a particular point we find that the coordinate basis may be changed by performing a transformation on the basis, leading to new coordinates derived from the old coordinates. Transformations of coordinates take a well-known form, which we will briefly recap. Suppose the two coordinate systems have basis vectors $x^1, \ldots, x^n$ and $y^1, \ldots, y^n$. Then for a given vector $\vec{V}$ with components u_i and v_j in the two coordinate systems it must be true that $u_i x^i = \vec{V} = v_j y^j$. Differentiating with respect to y, the relationship between coordinate systems is given by

$$v_j = u_i \frac{\partial x^i}{\partial y^j} \, . \tag{2.16}$$

This tells us how to find the components of a vector in a "new" coordinate system (the y-basis), given the components in the "old" coordinate system (the x-basis). Let us write $J^i_j = \partial x^i / \partial y^j$, and then since a summation is implied over i the transformation of coordinates can be written in terms of a matrix acting upon the components of vectors, $v_j = J^i_j u_i$. Such a matrix, relating two coordinate systems is called a *Jacobian matrix.* While one transformation matrix is needed to act upon vectors (which have only a single index), one transformation matrix per index is needed for more complex objects, e.g.

$$v_{ijk} = J^p_i J^r_j J^s_k u_{prs} \, . \tag{2.17}$$

Since the metric defines angles and lengths (and hence areas and volumes) calculations involving the volume of a region of spacetime (e.g. integration of a Lagrangian) must introduce a supplementary factor of $\sqrt{-g}$ (where g is the determinant of the metric $g^{\mu\nu}$) in order to remain invariant under arbitrary coordinate transformations. Hence instead of $d^n x \to d^n y$ we have

$$d^n x \sqrt{-g(x)} \to d^n y \sqrt{-g(y)} \, . \tag{2.18}$$

The square root of the determinant of the metric is an important factor in defining areas, as we can see by considering a parallelogram whose sides are defined by two vectors, $\vec{x}$ and $\vec{y}$ which are at an angle θ to each other. The area of this parallelogram is given by the magnitude of the cross product of these vectors, hence

$$
\begin{aligned}
A_{\text{parallel}} &= \sqrt{(\vec{x} \times \vec{y}) \cdot (\vec{x} \times \vec{y})} \\
&= \sqrt{(\vec{x} \cdot \vec{x})(\vec{y} \cdot \vec{y}) \sin^2 \theta} \\
&= \sqrt{(\vec{x} \cdot \vec{x})(\vec{y} \cdot \vec{y})(1 - \cos^2 \theta)} \, .
\end{aligned}
\tag{2.19}
$$

Suppose that $\vec{x}$ and $\vec{y}$ are basis vectors lying in a plane. Then the metric in this plane will be

$$m_{jk} = \begin{pmatrix} \vec{x} \cdot \vec{x} & \vec{x} \cdot \vec{y} \\ \vec{y} \cdot \vec{x} & \vec{y} \cdot \vec{y} \end{pmatrix} \tag{2.20}$$

where $j, k \in \{\vec{x}, \vec{y}\}$. Comparing Eqs. (2.19) and (2.20), we see that

$$A_{\text{parallel}} = \sqrt{(\vec{x} \cdot \vec{x})(\vec{y} \cdot \vec{y}) - (\vec{x} \cdot \vec{y})^2} = \sqrt{\det m_{ab}} \, . \qquad (2.21)$$

It is therefore reasonable to expect an analogous function of the metric to play a role in changes of coordinates. Furthermore we would expect the total area of some two-dimensional surface, which can be broken up into many small parallelograms, to be given by integrating the areas of such parallelograms together (this point will be taken up again in Sect. 6.3).

To see why Eq. (2.18) applies in the case of coordinate transformations,[5] consider an infinitessimal region of a space. Let this region be a parallelepiped in some coordinate system $x^1, \ldots, x^n$. Now suppose we want to change to a different set of coordinates, $y^1, \ldots, y^n$, which are functions of the first set (e.g. we want to change from polar coordinates to cartesian). The Jacobian of this transformation is

$$J = \frac{\partial(x^1, \ldots, x^n)}{\partial(y^1, \ldots, y^n)} = \begin{pmatrix} \frac{\partial x^1}{\partial y^1} & \cdots & \frac{\partial x^1}{\partial y^n} \\ \vdots & & \vdots \\ \frac{\partial x^n}{\partial y^1} & \cdots & \frac{\partial x^n}{\partial y^n} \end{pmatrix} \qquad (2.22)$$

The entries in the Jacobian matrix are the elements of the vectors defining the sides of the infinitessimal region we began with, referred to the new basis. Each row corresponds with one vector, and the absolute value of the determinant of such a matrix, multiplied by $d^n y = dy^1 \ldots dy^n$ gives the volume of the infinitessimal region. An integral referred to these new coordinates must include a factor of this volume, to ensure that the coordinates have been transformed correctly and the integral doesn't over-count the infinitessimal regions of which it is composed, hence

$$\int f(x^1, \ldots, x^n) dx^1 \ldots dx^n = \int f(y^1, \ldots, y^n) |\det J| dy^1 \ldots dy^n \, . \qquad (2.23)$$

The Jacobian matrix defines the transformation between coordinate systems. To be specific, we will choose the Minkowski metric $\eta_{\mu\nu}$ for the first coordinate system. The metric of the second coordinate system remains unspecified, hence

$$g_{\alpha\beta} = \frac{\partial x^\mu}{\partial y^\alpha} \frac{\partial x^\nu}{\partial y^\beta} \eta_{\mu\nu} \, . \qquad (2.24)$$

We can treat this expression as a product of matrices. If we do so, we must be careful about the ordering of terms, since matrix multiplication is non-commutative, and it is useful to replace one of the Jacobian matrices by its transpose. However this extra complication can be avoided since we are interested in the determinants of

---

[5] This argument is taken from Chap. 8 of [2], where a more detailed discussion can be found.

the matrices, and $\det(AB) = \det A \det B = \det B \det A$, and also $\det A^T = \det A$ so the ordering of terms is ultimately unimportant. Taking the absolute value of the determinant of Eq. (2.24),

$$|J| = \sqrt{\frac{g}{\eta}} = \sqrt{-g} \qquad (2.25)$$

since $\eta = \det \eta_{\mu\nu} = -1$. From this and Eq. (2.23) the use of a factor $\sqrt{-g}$ follows immediately.

The transformations described above, where a new coordinate basis is derived from an old one is called a *passive* transformation. By contrast, it is possible to leave the coordinate basis unchanged and instead change the positions of objects, whose coordinates will consequently change as measured in this basis. This is called an *active* coordinate transformation. With this distinction in mind, we will elaborate on the concept of diffeomorphism invariance in GR.

A diffeomorphism is a mapping of coordinates $f : x \to f(x)$ from a manifold $\mathcal{M}_1$ to a manifold $\mathcal{M}_2$ that is smooth, invertible, one-to-one, and onto. As a special case we can take $\mathcal{M}_1$ and $\mathcal{M}_2$ to be the same manifold, and define a diffeomorphism from a spacetime manifold to itself. A passive diffeomorphism will change the coordinates, but leave objects based on them unchanged, so that for instance the metric before a passive diffeomorphism is $g_{\mu\nu}(x)$ and after it is $g_{\mu\nu}(f(x))$. Invariance under passive diffeomorphisms is nothing special, as any physical theory can be made to yield the same results under a change of coordinates. An active diffeomorphism, on the other hand, would yield a new metric $g'_{\mu\nu}(x)$, which would in general measure different distances between any two points than does $g_{\mu\nu}(x)$. General relativity is significant for being invariant under active diffeomorphisms. This invariance requires that if $g_{\mu\nu}(x)$ is any solution of the Einstein field equations, an active diffeomorphism yields $g'_{\mu\nu}(x)$ which must be another valid solution of the EFEs. We require that any theory of quantum gravity should also embody a notion of diffeomorphism invariance, or at the very least, should exhibit a suitable notion of diffeomorphism invariance in the classical limit.

An understanding of classical general relativity helps us to better understand transformations between locally-defined coordinate systems. We will now proceed to a discussion of quantum field theory, where these local coordinate systems are abstracted to "internal" coordinates. And just as the discussion of GR provides us with tools to more easily visualise the concepts at the heart of QFT, the quantisation of field theories discussed in the next section will lay the foundations for our account of the attempts to extend classical GR into a quantum theory of gravity.

References

1. R.M. Wald, *General Relativity* (The University of Chicago Press, 1984). IISBN: 9780226870335. https://doi.org/10.7208/chicago/9780226870373.001.0001
2. D. Koks, *Explorations in Mathematical Physics: The Concepts Behind an Elegant Language* (Springer, 2006). ISBN: 978-0-387-30943-9. https://doi.org/10.1007/978-0-387-32793-8

Quantum Field Theory

3

Quantum Field Theory should be familiar to most (if not all) modern physicists, however we feel it is worth mentioning the basic details here, in order to emphasize the similarities between QFT and GR, and hence illustrate how GR can be written as a gauge theory. In short, we will see that a local change of phase of the wavefunction is equivalent to the position-dependent change of basis we considered in the case of GR. Just as the partial derivative of a vector gave (via the product rule) a derivative term corresponding to the change in basis, we will see that a derivative term arises corresponding to the change in phase of the quantum field. This introduces a connection and a covariant derivative defined in terms of the connection.

3.1 Covariant Derivative and Curvature

We may write the wavefunction of a particle as a product of wavefunctions $\phi(x)$ and $\omega(x)$ corresponding respectively to the external and internal degrees of freedom,[1]

$$\psi(x) = \phi(x)_j \omega(x)_j \tag{3.1}$$

where there is an obvious analogy to the definition of a vector, with the ω_j playing the role of basis vectors, the $\phi(x)_j$ playing the role of the components, and summation implied over the repeated index j. In complete analogy with Eq. (2.3), by applying the product rule we find that

$$\frac{d\psi}{dx^\mu} = \frac{\partial \phi_j}{\partial x^\mu} \omega_j + \phi_j \frac{\partial \omega_j}{\partial x^\mu}. \tag{3.2}$$

For illustrative purposes, let us consider a fairly simple choice of basis, where we have only one u and so we drop the index j. We will write $\omega = e^{ig\theta(x)}$. Then the derivative of ψ will take the form

---

[1] A more thorough discussion of the material in this subsection can be found in Chap. 3 of [1].

© Springer Nature Switzerland AG 2024
S. Bilson-Thompson, *Loop Quantum Gravity for the Bewildered*,
https://doi.org/10.1007/978-3-031-43452-5_3

$$\frac{d\psi}{dx^\mu} = \frac{\partial\phi}{\partial x^\mu}e^{ig\theta(x)} + ige^{ig\theta(x)}\phi\frac{\partial\theta(x)}{\partial x^\mu}$$

$$= e^{ig\theta(x)}\left(\frac{\partial}{\partial x^\mu} + ig\frac{\partial\theta(x)}{\partial x^\mu}\right)\phi. \tag{3.3}$$

Next we can pre-multiply the whole expression by $e^{-ig\theta(x)}$ to eliminate the exponential term on the right hand side. This is equivalent to Eq. (2.4) where we extracted an expression for the derivative of the components using $e^i \cdot e_j = \delta^i_j$. Lastly we switch notation slightly to more closely resemble Eq. (2.5), and define the term in brackets to be a covariant derivative

$$D_\mu = \partial_\mu + igA_\mu \tag{3.4}$$

where $A_\mu = \partial_\mu\theta$, and D_μ satisfies all the properties required of a derivative operator (linearity, Leibniz's rule, etc.).

A transformation $\theta \to \theta' = \theta + \zeta$ will result in a transformation of the wavefunction $\psi \to \psi' = e^{ig\zeta}\psi$, and a transformation of the connection $A_\mu \to A'_\mu$. For brevity, let us write $G = e^{ig\zeta}$. We can find the transformation of A_μ from the requirement that $D'_\mu\psi' = D'_\mu G\psi = GD_\mu\psi$, which means that

$$(\partial_\mu + igA'_\mu)G\psi = G(\partial_\mu + igA_\mu)\psi$$
$$\therefore (\partial_\mu G)\psi + G\partial_\mu\psi + igA'_\mu G\psi = G\partial_\mu\psi + igGA_\mu\psi$$
$$\therefore (\partial_\mu G)\psi + igA'_\mu G\psi = igGA_\mu\psi$$
$$\therefore igA'_\mu G = igGA_\mu - (\partial_\mu G)$$
$$\therefore A'_\mu = GA_\mu G^{-1} + \frac{i}{g}(\partial_\mu G)G^{-1}. \tag{3.5}$$

Substituting in $G = e^{ig\zeta}$ we deduce that A_μ transforms as

$$A'_\mu = A_\mu - \partial_\mu\zeta. \tag{3.6}$$

Since we defined $A_\mu = \partial_\mu\theta$ above, the presence of a minus sign might be a bit surprising. Surely from the definition of A_μ we expect that $\partial\theta' = \partial\theta + \partial\zeta$. However what Eq. (3.6) is telling us is simply that when we locally change the basis of a wavefunction but leave the overall physics unchanged, the connection must change in an equal and opposite manner to compensate. This is akin to the concept of diffeomorphism invariance discussed in Sect. 2.3. In both GR and QFT there are two ways to change the local coordinate basis. The first is by moving from an initial position to a new position where the basis is defined differently. The second is by staying at one point and performing a transformation (a diffeomorphism in GR, a gauge transformation in QFT) to change the coordinate basis. In each case, we want the laws of physics to remain the same, despite any change to the chosen coordinate basis. We can see how this condition is enforced by the transformation of

the connection, Eq. (3.6), and the role of the covariant derivative in the action for a
Dirac field ψ of mass m;

$$S = \int d^4x \, \overline{\psi}(\mathrm{i}\hbar c\gamma^\mu \partial_\mu - mc^2)\psi \,. \tag{3.7}$$

A *global* gauge transformation corresponds to rotating ψ by a *constant* phase $\psi \to e^{\mathrm{i}g\lambda}\psi$. Under this change we can see that the value of the action

$$S \to \int d^4x \, \overline{\psi}e^{-\mathrm{i}g\zeta}(\mathrm{i}\hbar c\gamma^\mu \partial_\mu - mc^2)e^{\mathrm{i}g\zeta}\psi \tag{3.8}$$

does not change because the factor of $e^{\mathrm{i}g\zeta}$ acting on ψ and the corresponding factor
of $e^{-\mathrm{i}g\zeta}$ acting on $\overline{\psi}$ pass through the partial derivative unaffected, and cancel out.
However if we allow ζ to become a function of position $\zeta(x)$, then the global gauge
transformation is promoted to a *local* gauge transformation, due to which the partial
derivative becomes

$$\partial_\mu \left(e^{\mathrm{i}g\zeta(x)}\psi\right) = e^{\mathrm{i}g\zeta(x)}\left(\partial_\mu + \mathrm{i}g(\partial_\mu\zeta(x))\right)\psi \tag{3.9}$$

leading to a modification of the action $S \to S - \int d^4x\,\hbar c\gamma^\mu(\partial_\mu\zeta)\overline{\psi}\psi$. The covariant
derivative, however, compensates for the x-dependence of ζ, since as we saw in
Eq. (3.5) it has the property that

$$D_\mu\psi \to D_\mu\left(e^{\mathrm{i}g\zeta(x)}\psi\right) = e^{\mathrm{i}g\zeta(x)}D_\mu\psi \tag{3.10}$$

and so the phase factor passes through the covariant derivative as desired. It is now
trivial to show that the Dirac action defined in terms of the covariant derivative,

$$S_{\mathrm{Dirac}} = \int d^4x \, \overline{\psi}(\mathrm{i}\hbar c\gamma^\mu D_\mu - mc^2)\psi \tag{3.11}$$

is invariant under local phase transformations of the form $\psi \to e^{\mathrm{i}g\zeta(x)}\psi$,
$\overline{\psi} \to \overline{\psi}e^{-\mathrm{i}g\zeta(x)}$, so long as $A_\mu(x)$ transforms as per Eq. (3.6). The connection A_μ
tells us how the phase of the wavefunction at each point corresponds to the phase at
a different point, in analogy to the connection in GR which told us how coordinate
bases varied from point to point, but additionally the requirement that the action
be invariant under *local* gauge transformations necessitates that it is not simply the
wavefunction, but also the connection that changes under a gauge transformation.

The discussion above has been restricted to the case of a simple rotation of the
phase (that is, $e^{\mathrm{i}g\zeta} \in U(1)$, the rotation group of the plane). In GR, by contrast,
the local bases at different points may be rotated in three dimensions relative to
each other (that is, the basis vectors are acted upon by elements of SO(3)). We can
accordingly generalise the discussion above to include phase rotations arising from

more elaborate groups. For instance, in the case of SU(2) we replace the wavefunction ψ by a Dirac doublet

$$\psi \rightarrow \psi = \begin{pmatrix} \psi_1(x) \\ \psi_2(x) \end{pmatrix} \tag{3.12}$$

and act upon this with transformations of the form

$$U(x) = \exp(i\zeta^I(x)t^I). \tag{3.13}$$

Here $t^I = \sigma^I/2$, (with σ^I the Ith Pauli matrix).[2] In this case the covariant derivative becomes

$$D_\mu = \partial_\mu + igA_\mu^I t^I \tag{3.14}$$

(summation on the repeated index is implied). In analogy to the case discussed above for GR, we can form the commutator of covariant derivatives. In this case, we obtain the field strength tensor $F_{\mu\nu}$, the analogue of the Riemann curvature tensor,

$$[D_\mu, D_\nu] = igF_{\mu\nu}^I t^I \tag{3.15}$$

where we can see (by applying the standard commutation relations for the Pauli matrices, namely $[\sigma^I, \sigma^J] = 2i\epsilon^{IJK}\sigma^K$, and relabelling some dummy indices) that

$$F_{\mu\nu}^I = \partial_\mu A_\nu^I - \partial_\nu A_\mu^I - g\epsilon^{IJK} A_\mu^J A_\nu^K. \tag{3.16}$$

When our gauge group is abelian (as in QED) all the generators of the corresponding Lie algebra commute with each other and thus the structure constants of the group (ϵ^{IJK} in the SU(2) example of Eq. (3.16)) vanish. In this event the field strength simplifies to

$$F_{\mu\nu}^I = \partial_\mu A_\nu^I - \partial_\nu A_\mu^I. \tag{3.17}$$

The field strength $F_{\mu\nu}^I$ itself is gauge *covariant* but not gauge *invariant*. Under an infinitesimal gauge transformation $A_0 \rightarrow A_0 + \delta A$ the field strength also changes by $F[A_0] \rightarrow F[A_0 + \delta A] = F_0 + \delta F$ where the variation in field strength is given by $\delta F = D_\mu[A_0]$ as the reader can easily verify by substituting and expanding in Eq. (3.16) or Eq. (3.17). Here $D_\mu[A_0]$ denotes that the covariant derivative is taken with respect to the original connection A_0.

The basic statement of Einstein's gravitational theory, often expressed in the saying

Matter tells geometry how to curve and geometry tells matter how to move.

---

[2] In general the t^I will be the appropriate generators of the symmetry group, e.g. $I = 1, 2, \ldots$ $N^2 - 1$ for SU(N).

has a parallel statement in the language of gauge theory. In a gauge theory, matter is represented by the fields ψ whereas the "geometry" (not of the background spacetime, but of the interactions between the particles) is determined by the configurations of the gauge field. The core idea of GR can then be generalised to an equivalent idea in field theoretic terms,

> Gauge charges tell gauge fields how to curve and gauge fields tell gauge charges how to move.

Now, what we have so far is an action, Eq. (3.11) which describes the dynamics of spinorial fields, interactions between which are mediated by the gauge field. The gauge field itself is not yet a dynamic quantity. In any gauge theory, consistency demands that the final action should also include terms which describe the dynamics of the gauge field alone. We know this to be true from our experience with QED where the gauge field becomes a particle called the photon. From classical electrodynamics Maxwell's equations possess propagating solutions of the gauge field - or more simply *electromagnetic waves*. The term giving the dynamics of the gauge field can be uniquely determined from the requirement of gauge invariance. We need to construct out of the field strength an expression with no indices. This can be achieved by contracting $F^I_{\mu\nu}$ with itself and then taking the trace over the Lie algebra indices. Doing this we get the term

$$S_{\text{gauge}} = -\frac{1}{4} \int d^4 x \, \text{Tr} \left(F^{\mu\nu} F_{\mu\nu} \right) \tag{3.18}$$

which in combination with (3.11) gives us the complete action for a gauge field interacting with matter

$$S = S_{\text{gauge}} + S_{\text{Dirac}} = \int d^4 x \left\{ -\frac{1}{4} \text{Tr} \left(F^{\mu\nu} F_{\mu\nu} \right) + \overline{\psi}(i\hbar c \gamma^\mu D_\mu - mc^2)\psi \right\}. \tag{3.19}$$

3.2 Dual Tensors, Bivectors and k-forms

The field strength is usually first encountered in the case of electromagnetism, where the relevant gauge group is U(1) which has only one group generator and so we can drop the index I in Eq. (3.17). Bearing in mind that we are using the convention $\eta_{\mu\nu} = \text{diag}(-1, +1, +1, +1)$, the electromagnetic field strength $F_{\mu\nu}$ combines the electric and magnetic fields into a single entity of the form

$$F_{\mu\nu} = \partial_\mu A_\nu - \partial_\nu A_\mu = \begin{pmatrix} 0 & -E_1 & -E_2 & -E_3 \\ E_1 & 0 & B_3 & -B_2 \\ E_2 & -B_3 & 0 & B_1 \\ E_3 & B_2 & -B_1 & 0 \end{pmatrix}. \tag{3.20}$$

Since each component of $F_{\mu\nu}$ is associated with two index values, we can think of the components as "bivectors" (oriented areas lying in the μ-ν plane), in analogy

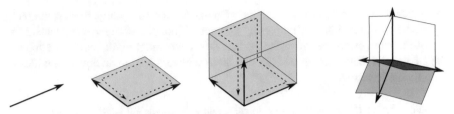

Fig. 3.1 Wedge products of basis vectors define basis bivectors, basis trivectors, and so on. While a vector's magnitude is its length, a bivector's magnitude is its area, and the magnitude of a trivector is its volume. The orientation of the unit bivector and unit trivector are shown here by the dashed arrows. The field strength $F_{\mu\nu}$ can be represented as a set of bivectors oriented between pairs of timelike and spacelike axes in four dimensions (shown here by distorting the angles between axes, as is done in a two-dimensional drawing of a cube). Shaded (unshaded) bivectors in the rightmost diagram are the magnetic (electric) field components

with vectors which carry only a single index (and are oriented lengths lying along a single axis). For the reader unfamiliar with bivectors we will very quickly review them.

A unit basis vector e_l can be visualised as a line segment with a "tail" and a "head", and an orientation given by traversing the vector from its tail to its head. A general vector is a linear combination of basis vectors, $\vec{v} = v^1 e_1 + v^2 e_2 + v^3 e_3 + \dots$ where the v^l are scalars. Similarly a unit basis bivector can be visualised as an area bounded by the vectors e_l and e_m, written as the wedge product $e_l \wedge e_m$, and with an orientation defined by traversing the boundary of this area along the first side, in the same direction as e_l, then along the second side parallel to e_m, and continuing anti-parallel to e_l and e_m to arrive back at the origin (this concept can be extended arbitrarily to define trivectors, etc. as illustrated in Fig. 3.1. Such a construction wedging k vectors together is called a k-blade[3] and a linear combination of k-blades, for several values of k, is called a multivector).

A general bivector is a linear combination of basis bivectors. Writing the field strength as such a general bivector we find that it takes the form

$$F_{\mu\nu} = E_1(e_1 \wedge e_0) + E_2(e_2 \wedge e_0) + E_3(e_3 \wedge e_0)$$
$$+ B_1(e_2 \wedge e_3) + B_2(e_3 \wedge e_1) + B_3(e_1 \wedge e_2). \tag{3.21}$$

Notice that the ordering of indices on the wedge products $(e_\mu \wedge e_\nu)$ has been arranged to match the μth row and νth column of Eq. (3.20). Electric fields are those parts of $F_{\mu\nu}$ lying in a plane defined by one space axis and the time axis, while magnetic fields are those lying in a plane defined by two space axes (Fig. 3.1). Reversing the orientation of a bivector is equivalent to traversing its boundary "backwards", so we may write $e_\mu \wedge e_\nu = -e_\nu \wedge e_\mu$. This is consistent with the fact that the field strength is antisymmetric, i.e. $F_{\mu\nu} = -F_{\nu\mu}$.

---

[3] The term k-vector is also used, though we feel it is best avoided as it could lead to confusion with the concept of a vector in k dimensions.

We can also combine the electric and magnetic fields into a single entity by defining the dual field strength,

$$\star F_{\mu\nu} = \frac{1}{2}\epsilon_{\mu\nu}{}^{\lambda\rho}F_{\lambda\rho} = \begin{pmatrix} 0 & B_1 & B_2 & B_3 \\ -B_1 & 0 & E_3 & -E_2 \\ -B_2 & -E_3 & 0 & E_1 \\ -B_3 & E_2 & -E_1 & 0 \end{pmatrix}, \tag{3.22}$$

where we can of course raise or lower indices on the field strength tensor and antisymmetric tensor using the metric. We can see that the mapping between field strength and dual field strength[4] associates a given electric field component with a corresponding magnetic field component, such that $E_i \to -B_i$ and $B_j \to E_j$. Thinking in terms of bivectors, the quantity defined on the plane between any pair of spacetime axes is associated to the quantity defined on the plane between the other two spacetime axes. The field strength is said to be *self-dual* if $\star F = +F$, and *anti-self-dual* if $\star F = -F$. Although we will not be concerned with (anti-)self-dual field strengths in the rest of this book, we will be dealing with (anti-)self-dual gauge connections from Sect. 5.1 onwards. The EM field strength as presented here is merely the simplest example to use to introduce the concept of self-duality, and illustrate its physical meaning. Further discussion of duality, for the reader requiring a deeper understanding, is presented in the appendices, Sect. B.3. Some readers will also no doubt have noticed the similarity between bivectors $e_l \wedge e_m$, and differential 2-forms $dx^l \wedge dx^m$. The two are indeed very similar.

A bivector defined by the wedge product of two vectors $\vec{u} \wedge \vec{v}$ can be imagined as a parallelogram with two sides parallel to $\vec{u}$, and the other sides parallel to $\vec{v}$. The magnitude of this bivector is the area of the enclosed parallelogram.[5] Differential forms, on the other hand, have a magnitude which is thought of as a density. This is often drawn as a series of lines (similar to the contour lines on a topographical map or the isobars on a weather map) with smaller spacing between lines indicating higher density (Fig. 3.2). Hence a 1-form can be thought of as a density of contour lines or contour surfaces perpendicular to the direction of the 1-form. The inner product of a vector with a 1-form is a scalar - the number of lines that the vector crosses. Similarly a 2-form can be thought of as a series of contours spreading out through a plane (this plane being defined by the directions of the two 1-forms wedged together to make the 2-form). Clearly there is a one-to-one mapping between vectors and 1-forms, and between bivectors and 2-forms, which simply involves changing one's choice of magnitude, (length or area) $\leftrightarrow$ (density).

It is certainly more common to see 1-forms, 2-forms, and other higher-dimensional forms used throughout physics, but k-blades and multivectors can be very useful too, and are often easier to visualise (see Appendix B for a further discussion of

---

[4] The notation $\tilde{F}$ is also used for the dual field strength.

[5] Bivectors could just as well be visualised as disks or ellipses or any other planar shape. The area and orientation are what really matter. But describing them as parallelograms helps motivate the process of building up bivectors, trivectors, etc. by wedging together several vectors.

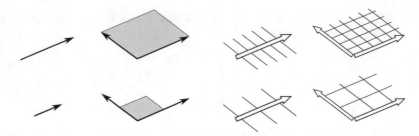

Fig. 3.2 Two vectors (far left) with the same direction and different magnitudes differ in length, while two bivectors (left) differ in their areas. The magnitude of k-forms is a density, and can be represented by interval lines. A 1-form (right) has a direction (indicated by the hollow arrows), just like a vector, but the spacing of interval lines represents its magnitude. A 2-form (far right) defines a plane, just like a bivector does, and once again the magnitude is represented by the spacing of interval lines. In all cases, the greater magnitude object is on the top row

the relationship between k-blades and differential forms). The use of k-blades in preference to differential forms has been extensively developed by Hestenes [5].

3.3   Wilson Loops and Holonomies

In Chap. 2 we defined a loop holonomy as a measure of how much the initial and final values of a vector transported around a closed loop differ. The discussion in the previous section demonstrates that the internal degrees of freedom of a spinor can also be position-dependant, and hence it should be possible to define a loop holonomy by the difference between the initial and final values of a spinor transported around a closed loop.[6] As a first step to constructing such a definition, let us consider what happens when we compare the values of a field at different points, separated by a displacement dx^μ. We begin by using Eqs. (3.2), (3.3) and (3.4) to write

$$\frac{d\psi}{dx^\mu} = \frac{\partial\phi}{\partial x^\mu}\omega + \phi\frac{\partial\omega}{\partial x^\mu} = \omega\left(\partial_\mu + \mathbf{i}gA_\mu\right)\phi \tag{3.23}$$

from which by comparing terms we readily see that $\mathbf{i}gA_\mu\omega = \partial_\mu\omega$, or equivalently $\mathbf{i}gA_\mu\omega dx^\mu = d\omega$. The internal components of the fields will be related by a gauge rotation which we will call $U(dx^\mu)$. The action of this rotation can be expanded as

$$U(dx^\mu)\omega = \omega + d\omega = \omega + \mathbf{i}gA_\mu\omega dx^\mu = (1 + \mathbf{i}gA_\mu dx^\mu)\omega \tag{3.24}$$

and we immediately see that

$$U(dx^\mu) = \exp\{\mathbf{i}gA_\mu dx^\mu\}. \tag{3.25}$$

---

[6] The name holonomy is also used within the LQG community to refer to a loop or closed path. We feel this is unnecessarily confusing, hence our choice to use "loop holonomy" for a quantity *associated to* a closed path, and to avoid using the term "holonomy" for a closed path itself. The reader should be aware that this terminology does, however, exist within the wider literature.

$U(dx^\mu)$ is the parallel transport operator that allows us to bring two field values at different positions together so that they may be compared. Remembering that the effect of parallel transport is path-dependant, this operator can be readily generalised to finite separations along an arbitrary path λ and connections valued in arbitrary gauge groups, in which case we find

$$U(x, y) = \mathcal{P} \exp \left\{ \int_\lambda igA_\mu{}^I(x)t^I dx^\mu \right\}$$ (3.26)

where the $\mathcal{P}$ tells us that the integral must be *path ordered*,[7] t^I are gauge group generators as before, and x and y are the two endpoints of the path λ we are parallel transporting along. If the gauge connection vanishes along this path then the gauge rotation is simply the identity matrix and ψ is unchanged by being parallel transported along the path. In general, however, the connection will *not* vanish.

Now consider the situation when the path λ is a closed loop, i.e. its beginning and end-point coincide. Analogously to the situation for a curved manifold, where the parallel transport of a vector along a closed path gives us a measure of the curvature of the spacetime bounded by that path, the parallel transport of a spinor around a closed path yields a measure of the *gauge* curvature living on a surface bounded by this path. We can see this simply in the case of a small square "plaquette" in the μ-ν plane, with side length a. The gauge rotation in this case is a product of the rotation induced by parallel-transporting a spinor along each of the four sides of the plaquette in order. The parallel transport operators for each side of the plaquette are found from Eq. (3.26), and explicitly, their product around a plaquette is

$$W = e^{igaA_\nu^\dagger(x+a\nu)} e^{igaA_\mu^\dagger(x+a\mu+a\nu)} e^{igaA_\nu(x+a\mu)} e^{igaA_\mu(x)} .$$ (3.27)

Assuming that we are dealing with a non-Abelian field theory, this product of exponentials can be converted to a single exponential by use of the Baker-Cambell-Haussdorf rule, which for the product of four terms takes the form

$$e^A e^B e^C e^D = \exp\{A + B + C + D + [A, B] + [A, C] + [A, D] + [B, C] + [B, D] + [C, D] + ...\} .$$ (3.28)

After a bit of algebra we find that this simplifies to

$$W = \exp\{iga^2 F_{\mu\nu} + ...\}$$ (3.29)

where the ... represent higher-order terms. An arbitrary loop can be approximated by a tiling of small plaquettes, to yield a result proportional to the total tiled area multiplied by $F_{\mu\nu}$. Since the common edges of adjacent plaquettes are traversed in opposite directions the contributions along these edges are cancelled, and the entire tiling results in a path around the outside of the tiled area (Fig. 3.3). Such an arbitrary

---

[7] See Appendix C for the definition of a "path ordered" exponential.

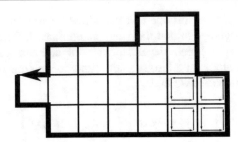

Fig. 3.3 An arbitrary closed path in the plane can be approximated by tilings of plaquettes. Since each plaquette is traversed anti-clockwise, adjacent edges make cancelling contributions to the parallel transport of a spinor, leaving only the contribution at the boundary of the tiling (as illustrated for the plaquettes in the lower-right corner)

loop is called a Wilson loop, and the loop holonomy associated to it is called the *Wilson loop variable*, and corresponds to an element in the gauge group of the theory. To obtain a single variable from the parallel transport around a loop, we take the trace of the loop holonomy

$$W_\lambda = \text{Tr}\mathcal{P} \exp\left\{ \oint_\lambda igA_\mu^I(x)t^I dx^\mu \right\}. \tag{3.30}$$

The Wilson loop is gauge-invariant, since each line segment of which the loop is composed transforms as

$$U(x, y) \rightarrow G(y)U(x, y)G^{-1}(x) \tag{3.31}$$

under a gauge transformation like that in Eq. (3.5), and so the product of several line segments forming a closed loop transforms as

$$W \rightarrow W' = G(x_1)U(x_1, x_2)G^{-1}(x_2)\dots G(x_n)U(x_n, x_1)G^{-1}(x_1). \tag{3.32}$$

Different gauge transformations therefore correspond with different choices of starting point for the loop. However the trace is invariant under cyclic permutations, $\text{Tr}ABC = \text{Tr}BCA = \text{Tr}CAB$, and so the Wilson loop variable is independent of choice of gauge transformation [2]. It is a fairly straightforward matter to see from the form of Eq. (3.18) that since W is the exponential of a term proportional to $F_{\mu\nu}$ the action for the gauge field may be constructed by evaluating Wilson loops.

This discussion shows that $F_{\mu\nu}$ is a measure of the gauge curvature within a surface, as well as a measure of the loop holonomy of the loop enclosing the surface (that is, the gauge rotation induced on a spinor when it is parallel-transported around a closed loop). Hence when the connection does not vanish the associated loop holonomy will in general not be trivial.

3.4 Dynamics of Quantum Fields

We will conclude this chapter with a discussion of two approaches to the dynamics of quantum fields. These are well-established in the case of theories like QED and QCD, and so it will be natural later on to consider equivalent approaches when we wish to quantise spacetime, which is the dynamical field in GR. These two approaches are based on Lagrangian and Hamiltonian dynamics.

Lagrangian (or Path Integral) Approach

Although our primary concern is fields, some insight can be provided by considering the classical case of a non-relativistic point particle in flat space moving under the influence of an external potential $V(x)$ for which the action is given by

$$S_{pp}[\lambda] = \int_\lambda d^3x dt \left(\frac{1}{2}m\dot{x}^2 - V(x)\right) = \int_\lambda d^3x dt \, L(x). \tag{3.33}$$

The action integral depends on the choice of the path λ taken by the particle as it moves between its initial to final positions. By considering infinitesimal variations of the path we find that the classical behaviour of the particle corresponds to following a path for which the action is an extremum (generally a minimum). This amounts to finding the equation(s) of motion by inserting the Lagrangian L into the Euler-Lagrange equations.

Note that in the relativistic case the potential term must be replaced by a gauge field A_μ in which case the action takes the form

$$S_{rel}[\lambda] = \int_\lambda d^3x dt \, \frac{(p^\mu + A^\mu)(p_\mu + A_\mu)}{m_0} \tag{3.34}$$

where p^μ is the energy-momentum 4-vector of the particle and m_0 is its rest mass. This is the familiar action for a charged point particle moving under the influence of an external potential encoded in the Abelian gauge potential A_μ.

In the path-integral approach to quantum mechanics we assign a complex amplitude[8] (or real probability in the Euclidean case) to any path λ by

$$\exp\{iS[\lambda]\}. \tag{3.35}$$

---

[8] The reader is reminded that by performing a Wick rotation, to introduce an imaginary time parameter, a mapping can be established between path integrals and statistical physics. See e.g. Chap. 25 of [3].

Fig. 3.4 The path a particle takes between initial and final positions can be regarded as a series of paths between intermediate positions. Varying these intermediate positions leads to variations in the overall path, and the action associated to it

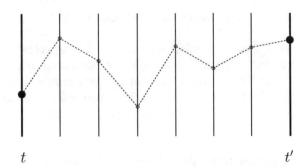

t t'

In contrast to the classical case we do not restrict our attention to those paths which correspond to extrema of the action. Instead we use this complex amplitude as a weighting function which allows us to calculate matrix elements for transitions between an arbitrary pair of initial and final states by forming a weighted sum of all paths which interpolate between the two states.

For the point-particle $|q, t\rangle$ represents a state where the particle is localized at position q at time t. The path between initial and final states can be broken up into a series of small steps by slicing up the time interval $t' - t$, as depicted in Fig. 3.4. The path is varied by adjusting the positions the particle passes through at the intermediate times t_1, t_2, etc. just as we considered variations of the path of a particle in the classical case. Each variation results in an action associated to the corresponding path. The matrix element between states at two different times then takes the form

$$\langle q', t' \mid q, t \rangle = \int \mathcal{D}[\psi] \exp\left\{ \mathrm{i}S_{\mathrm{pp}}[\lambda] \right\} . \tag{3.36}$$

The weighting factor gives higher value to the contribution from those paths which have an associated action close to the minimum. It is this which results in classical behaviour, in the appropriate limit. However the contributions of all possible paths must still be taken into account to accurately calculate the transitions between states.

We now turn from considering single particles, to the path-integral approach to quantum field theory. Here the basic element is the propagator (or the partition function when the space on which the field is defined is a Euclidean manifold) which allows us to calculate the probability amplitudes between pairs of initial and final states of our Hilbert space. As shown in Eq. (3.18), starting with the curvature of a gauge field it is possible to define an action which governs the dynamics of the gauge field. Thus an analogous approach to the path integral just described for particles can be adopted when we wish to consider fields. In this case we do not sum over paths, but field configurations in each possible *history* between initial and final states,

$$\langle \Psi_i(t) \mid \Psi_f(t') \rangle = \int \mathcal{D}[\psi] \exp\left\{ \mathrm{i}S[\lambda] \right\}, \tag{3.37}$$

in contrast to the classical view of dynamics, in which a system moves from an initial state to a final state in exactly one way. Here $\mathcal{D}[\psi]$ is an appropriate measure on the space of allowed field configurations. Of course, as is well known, quantum field theory allows for the creation and annihilation of particle-antiparticle pairs. Hence the transition between an initial and final state is not so straightforward as the sum over contributions from all possible paths a single particle could take. The notational simplicity of writing $S[\lambda]$ (which emphasises the conceptual similarity between the dynamics of single particles, quantum fields, and as we shall soon see, spacetime itself) in this case conceals the presence of interaction and source terms within the exponential. Taylor expanding the contributions of these leads to a perturbation series of contributions to the overall calculation, which are identified with Feynman diagrams for interactions involving various numbers of incoming, outgoing, and intermediate states (see e.g. Sect. I.7 of [4], or Chap. 24 of [3]).

Hamiltonian Approach: Canonical Quantisation

The alternative to the Lagrangian or path-integral approach is to study the dynamics of a system through its Hamiltonian. This leads to Dirac's procedure for canonical (or "second") quantisation.[9] The Hamiltonian H for a dynamical system can be constructed from the Lagrangian L by performing a Legendre transformation. Given a configuration variable q, which we can think of as a generalised position, and a corresponding generalised momentum p defined by

$$p = \frac{\partial L}{\partial \dot{q}},\tag{3.38}$$

then the Hamiltonian is given by

$$H[p, q] = p\dot{q} - L[q, \dot{q}]\tag{3.39}$$

in the case of a point particle, and generalisations of this equation for other systems. If we define the Poisson bracket of two functions by

$$\{f, g\} = \sum_{i=1}^{n} \left(\frac{\partial f}{\partial q_i} \frac{\partial g}{\partial p_i} - \frac{\partial f}{\partial p_i} \frac{\partial g}{\partial q_i} \right)\tag{3.40}$$

where $f = f(q, p, t)$ and $g = g(q, p, t)$, then Hamilton's equations can be written in the form

$$\dot{q} = \frac{\partial H}{\partial p} = \{H, p\} \quad \text{and} \quad \dot{p} = -\frac{\partial H}{\partial q} = \{H, q\}\tag{3.41}$$

---

[9] The quantisation of the motion of a particle in a classical potential is sometimes referred to as "first quantisation". This is the basis for the somewhat un-intuitive name "second quantisation" for quantisation extended to the potential as well.

and give the time evolution of the system. Hence, leaving the second spot in the brackets empty, time evolution is generated by the operator $\{H, \}$ which acts upon the generalised coordinates and momenta.

In quantum mechanics and quantum field theory observables are replaced by operators, i.e. $x \rightarrow \hat{x}$. While operators do not necessarily commute, classical observables do. However the Poisson bracket of two observables will not necessarily be zero, and Dirac was led to postulate that in the transition from classical to quantum mechanics, Poisson brackets between observables should be replaced by commutation relations, where the scalar value of the commutator is $i\hbar$ times the scalar value of the equivalent Poisson bracket, i.e.

$$\{f, g\} = 1 \quad \text{implies} \quad \left[\hat{f}, \hat{g}\right] = i\hbar. \tag{3.42}$$

This prescription will be central to our attempts to quantise spacetime in later sections.

We are now ready to (at least attempt to) apply our understanding of quantum field theories to the formulation of general relativity.

References

1. K. Moriyasu, *An Elementary Primer for Gauge Theory*. (World Scientific, 1983). ISBN: 978-9971-950-94-1. https://doi.org/10.1142/0049
2. M.E. Peskin, D.V. Schroeder, *An Introduction to Quantum Field Theory*. (Addison Wesley, 1995). ISBN: 9780201503975. https://doi.org/10.1201/9780429503559
3. T. Lancaster, S.J. Blundell, *Quantum Field Theory for the Gifted Amateur*. (Oxford University Press, 2015). ISBN: 9780199699322. https://doi.org/10.1093/acprof:oso/9780199699322.001. 0001
4. A. Zee, *Quantum Field Theory in a Nutshell*, 2nd edn. (Princeton University Press, 2010). ISBN: 9780691140346. https://doi.org/10.1063/1.1752429
5. D. Hestenes, Space-Time Algebra, (Birkhauser, 2015). ISBN: 978-3-319-18413-5. https://doi. org/10.1007/978-3-319-18413-5

Expanding on Classical GR

4

We now return to the discussion of general relativity. Equipped with the preceding discussions of both the quantisation of field theories, and the geometrical interpretations of gauge transformations, it is time to set about formulating what will eventually become a theory of dynamical spacetime obeying rules adapted from quantum field theory. But before we get there we must cast classical GR into a form amenable to quantisation.

From classical mechanics, and the discussion in Sect. 3.4, we know that dynamics can be described either in the Hamiltonian or the Lagrangian frameworks. The benefits of a Lagrangian framework are that it provides us with a covariant perspective on the dynamics and connects with the path-integral approach to the quantum field theory of the given system. The Hamiltonian approach, on the other hand, provides us with a phase space picture and access to the Schrödinger method for quantization. Each has its advantages and difficulties and thus it is prudent to be familiar with how both frameworks may be applied to general relativity. We will begin by discussing these approaches in a classical framework, and move to quantisation in Chap. 5.

4.1 Lagrangian Approach: The Einstein-Hilbert Action

The form of the Lagrangian, and hence the action, can be determined by requirements of covariance and simplicity. Out of the dynamical elements of geometry—the metric and the connection—we can construct a limited number of quantities which are invariant under coordinate transformations, hence they should have no uncontracted indices. These quantities must be constructed out of the Riemann curvature tensor or its derivatives. These possibilities are of the form: $\{R, R_{\mu\nu}R^{\mu\nu}, R^2, \nabla_\mu R\nabla^\mu R, \ldots\}$. The simplest of these is the Ricci scalar $R = R_{\mu\nu\alpha\beta}g^{\mu\alpha}g^{\nu\beta}$. As it turns out this term is sufficient to fully describe Einstein's general relativity, yielding a Lagrangian that is simply $\sqrt{-g}R$, where as noted in Sect. 2.3, $g = \det(g^{\mu\nu})$.

© Springer Nature Switzerland AG 2024
S. Bilson-Thompson, *Loop Quantum Gravity for the Bewildered*,
https://doi.org/10.1007/978-3-031-43452-5_4

This allows us to construct the simplest Lagrangian which describes the coupling of geometry to matter,

$$S_{\text{EH+M}} = \frac{1}{\kappa} \int d^4x \sqrt{-g} R + \int d^4x \sqrt{-g} \mathcal{L}_{\text{matter}} \tag{4.1}$$

where $\mathcal{L}_{\text{matter}}$ is the Lagrangian for the matter fields that may be present and κ is a constant, to be determined. If the matter Lagrangian is omitted, one obtains the usual vacuum field equations of GR. This action (i.e. omitting the matter term) is known as the Einstein-Hilbert action, S_{EH}.

It is worth digressing to prove (at least in outline form) that the Einstein field equations can be found from $S_{\text{EH+M}}$. The variation of the action (4.1) yields a classical solution which, by the action principle, is chosen to be zero,

$$\delta S = 0 = \int d^4x \left[\frac{1}{\kappa} \frac{\delta \sqrt{-g}}{\delta g^{\mu\nu}} R + \frac{1}{\kappa} \sqrt{-g} \frac{\delta R}{\delta g^{\mu\nu}} + \frac{\delta \sqrt{-g} \mathcal{L}_{\text{matter}}}{\delta g^{\mu\nu}} \right] \delta g^{\mu\nu} \tag{4.2}$$

which implies that

$$\frac{1}{\sqrt{-g}} \frac{\delta \sqrt{-g}}{\delta g^{\mu\nu}} R + \frac{\delta R}{\delta g^{\mu\nu}} = -\kappa \frac{1}{\sqrt{-g}} \frac{\delta \sqrt{-g} \mathcal{L}_{\text{matter}}}{\delta g^{\mu\nu}} . \tag{4.3}$$

The energy-momentum tensor can be defined as

$$T^{\mu\nu} = -\frac{2}{\sqrt{-g}} \frac{\delta \sqrt{-g} \mathcal{L}_{\text{matter}}}{\delta g^{\mu\nu}} \tag{4.4}$$

where $g = \det(g^{\mu\nu})$, and $\mathcal{L}_{\text{matter}}$ is a Lagrangian encoding the presence of matter.[1] From Eq. (4.4) we can immediately see that

$$\frac{1}{\sqrt{-g}} \frac{\delta \sqrt{-g}}{\delta g^{\mu\nu}} R + \frac{\delta R}{\delta g^{\mu\nu}} = \frac{\kappa}{2} T^{\mu\nu} . \tag{4.5}$$

We now need to work out the variation of the terms on the left-hand-side. Omitting the details, which can be found elsewhere (see e.g. the appendix of [2]), we find that

$$\delta \sqrt{-g} = -\frac{1}{2\sqrt{-g}} \delta \sqrt{g} = \frac{1}{2} \sqrt{-g} (g^{\mu\nu} \delta g_{\mu\nu}) = -\frac{1}{2} \sqrt{-g} (g_{\mu\nu} \delta g^{\mu\nu}) \tag{4.6}$$

---

[1] This definition of the energy-momentum tensor may seem to come out of thin air, and in many texts it is simply presented as such. To save space we will follow suit, but the reader who wishes to delve deeper should consult [1], in which $T_{\mu\nu}$ is referred to as the *dynamical* energy-momentum tensor, and it is proven that it obeys the conservation law $\nabla_\mu T^{\mu\nu} = 0$ (as one would hope, since energy and momentum are conserved quantities), as well as being consistent with the form of the electromagnetic energy-momentum tensor.

thanks to Jacobi's formula for the derivative of a determinant. The variation of the Ricci scalar can be found by differentiating the Riemann tensor, and contracting on two indices to find the variation of the Ricci tensor. Then, since the Ricci scalar is given by $R = g^{\mu\nu} R_{\mu\nu}$ we find that

$$\delta R = R_{\mu\nu} \delta g^{\mu\nu} + g^{\mu\nu} \delta R_{\mu\nu} . \tag{4.7}$$

The second term on the right may be neglected when the variation of the metric vanishes at infinity, and we obtain $\delta R / \delta g^{\mu\nu} = R_{\mu\nu}$. Plugging these results into Eq. (4.5) we find that

$$-\frac{1}{2} g_{\mu\nu} R + R_{\mu\nu} = \frac{\kappa}{2} T^{\mu\nu} \tag{4.8}$$

which yields the Einstein equations if we set $\kappa = 16\pi\mathcal{G}$.

As noted in Eq. (2.11), we can write $\Gamma^{\rho}{}_{\mu\nu}$ in terms of the metric $g_{\mu\nu}$,

$$\Gamma^{\rho}{}_{\mu\nu} = \frac{1}{2} g^{\rho\delta} \left(\partial_\mu g_{\delta\nu} + \partial_\nu g_{\delta\mu} - \partial_\delta g_{\mu\nu} \right)$$

and since the covariant derivative ∇_μ is a function of $\Gamma^{\rho}{}_{\mu\nu}$, and the Riemann tensor is defined in terms of the covariant derivative, the Einstein-Hilbert action is ultimately a function of the metric $g_{\mu\nu}$ and its derivatives.

As a further aside, we will briefly describe how the Lagrangian formulation allows us to make contact with the path-integral or sum-over-histories approach outlined in Sect. 3.4, and apply it to the behaviour of spacetime as a dynamical field. In general, this approach involves calculating transition amplitudes with each path between the initial and final states being weighted by an exponential function of the action associated with that path. In the case of gravity we may think of four-dimensional spacetime as a series of spacelike hypersurfaces, Σ_t, corresponding to different times. Each complete 4-dimensional geometry consisting of a series of 3-dimensional hypersurfaces that interpolate between the initial and final states may be thought of as the generalisation of a "path". This $3 + 1$ splitting of spacetime into foliated three-dimensional hypersurfaces will be covered in more detail in the next section. To calculate the matrix-elements (as in Eq. (3.37)) for transition amplitudes between initial and final states of geometry, Σ_t and $\Sigma_{t'}$ (see Fig. 4.1) we use the Einstein-Hilbert action for GR on a manifold $\mathcal{M}$ without matter

$$S_{\text{EH}} = \frac{1}{\kappa} \int d^4 x \sqrt{-g} \, R. \tag{4.9}$$

Let us represent the states corresponding to the initial and final hypersurfaces as $|h_{ab}, t\rangle$ and $|h'_{ab}, t'\rangle$, where h_{ab} is the intrinsic metric[2] of a given spatial hypersurface,

---

[2] The intrinsic metric will be introduced properly very shortly, specifically in Eq. (4.16) and the associated discussion.

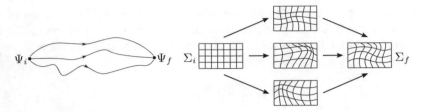

Fig. 4.1 Weighted sums of transitions between different configurations of spacelike hypersurfaces may be used to calculate the transition amplitude between an initial and final state of geometry (right), see Eq. (4.10). This is analogous to the path-integral approach used in quantum field theory (left), see Eq. (3.37)

and $a, b \in \{1, 2, 3\}$. Then the probability that evolving the geometry will lead to a transition between these two states is given by

$$\langle h_{ab}, t | h'_{ab}, t' \rangle = \int \mathcal{D}[g_{\mu\nu}] \exp\left\{ iS_{EH}(g_{\mu\nu}) \right\} \tag{4.10}$$

where the action is evaluated over all 4-metrics $g_{\mu\nu}$ interpolating between the initial and final hypersurfaces. $\mathcal{D}[g_{\mu\nu}]$ is the appropriate measure on the space of 4-metrics. While this approach is noteworthy, and ultimately leads to a very successful *computational* approach to quantising gravity [3], it is not the path we follow to formulate loop quantum gravity. Instead, as mentioned above, the Lagrangian formulation of general relativity is used as a stepping-stone to the Hamiltonian formulation.

4.2   Hamiltonian Approach: The ADM Splitting

Since general relativity is a theory of dynamical spacetime, we will want to describe the dynamics of spacetime in terms of some variables which make computations as tractable as possible. The Hamiltonian formulation is well suited to a wide range of physical systems, and the ADM (Arnowitt-Deser-Misner) formalism, described below, allows us to apply it to general relativity. We can think of the action, Eq. (4.1), which is clearly written in the form of an integral of a Lagrangian, as a stepping-stone to this Hamiltonian approach. This Hamiltonian formulation of GR takes us to the close of our discussion of classical gravity, and will be used as the jumping-off point for the quantisation of gravity, to be undertaken in Chap. 5.

The ADM formalism involves foliating spacetime into a set of three-dimensional spacelike hypersurfaces, and picking an ordering for these hypersurfaces which plays the role of time, so that the hypersurfaces are level surfaces of the parameter t. This is a necessary feature of the Hamiltonian formulation of a dynamical system, although it seems at odds with the way GR treats space and time as interchangable parts of spacetime. However this time direction is actually a "fiducial time"[3] and will turn

---

[3] The term "fiducial" refers to a standard of reference, as used in surveying, or a standard established on a basis of faith or trust.

Fig. 4.2 When performing
the ADM splitting, the lapse
function N and shift vector
N^μ define how points on
successive hypersurfaces are
mapped together

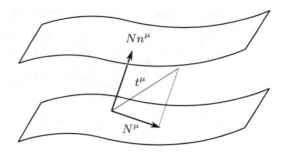

out not to affect the dynamics. It is essentially a parameter used as a scaffold, which
in the absence of a metric is not directly related to the passage of time as measured
by a clock.

To begin, we will suppose that the 4-dimensional spacetime is embedded within
a manifold $\mathcal{M}$ (which may be $\mathbb{R}^4$ or any other suitable manifold). Next we choose a
local foliation[4] $\{\Sigma_t, t\}$ of $\mathcal{M}$ into spacelike 3-manifolds, where Σ_t is the 3-manifold
corresponding to a given value of the parameter t. We will refer to such a manifold
as a "leaf of foliation". The topology of the original four-dimensional spacetime is
then $\Sigma \times \mathbb{R}$, while t is a parametrization of the set of geodesics orthogonal to Σ_t
(c.f. Fig. 4.2). In addition at each point of a leaf we have a unit time-like vector n^μ
(with $n^\mu n_\mu = -1$) which defines the normal at each point on the leaf.

Given the full four-metric $g_{\mu\nu}$ on $\mathcal{M}$ and the vector field n^μ the foliation is
completely determined by the requirement that the surfaces Σ_t of constant "time"
are normal to n^μ.

The diffeomorphism invariance of general relativity implies that there is no canonical choice of the time-like vector field t^μ which maps a point x^μ on a leaf Σ_t to
the point x'^μ on the leaf $\Sigma_{t+\delta t}$, i.e. which generates time evolution of the geometry.
This property is in fact the gauge symmetry of general relativity. It implies that we
can choose any vector field t^μ as long as it is time-like. Such a vector field can be
projected onto the three-manifold to obtain the *shift vector* $N^a = t_\parallel$ which is the part
tangent to the surface, while the component of t^μ normal to the three-manifold is then
identified as the "distance between hypersufaces" and is called the *lapse function*
$N = t_\perp$. Therefore t^μ can be written as

$$t^\mu = Nn^\mu + N^\mu \tag{4.11}$$

where, though we have written the shift as a four-vector to keep our choice of indices
consistent, it is understood that $N^0 = 0$ in a local basis of coordinates adapted to the
splitting.

---

[4] Generally one assumes that our 4-manifolds can always be foliated by a set of spacelike 3-
manifolds. For a general theory of quantum gravity the assumption of trivial topologies must be
dropped. In the presence of topological defects in the 4-manifold, in general, there will exist inequiv-
alent foliations in the vicinity of a given defect. This distinction can be disregarded in the following
discussion for the time being.

Now we can determine the components of the four-metric in a basis adapted to the splitting as follows;

$$
\begin{aligned}
g_{00} &= g_{\mu\nu} t^\mu t^\nu \\
&= g_{\mu\nu} \left(N n^\mu + N^\mu \right) \left(N n^\nu + N^\nu \right) \\
&= N^2 n^\mu n_\mu + N^\mu N_\mu + 2\, N (N^\mu n_\mu) \\
&= -N^2 + N^\mu N_\mu
\end{aligned}
\tag{4.12}
$$

where we have used $n^\mu n_\mu = -1$ and $N^\mu n_\mu = 0$ in the third line. Working in a coordinate basis where $N^\mu = (0, N^a)$, the result $g_{00} = -N^2 + N^a N_a$ is found.[5] Similarly to obtain the other components of the metric we project along the time-space and the space-space directions,

$$
g_{\mu\nu} t^\mu N^\nu = N^\mu N_\mu \equiv N^a N_a.
\tag{4.13}
$$

Since, by definition $g_{0\nu} \equiv g_{\mu\nu} t^\mu$, this implies that $g_{0a} = N_a$. The space-space components of $g_{\mu\nu}$ are simply given by selecting values of the indices μ, $\nu \in \{1, 2, 3\}$. Thus the full metric $g_{\mu\nu}$ can be written schematically as

$$
g_{\mu\nu} = \begin{pmatrix} -N^2 + N^a N_a & N \\ N^T & g_{ab} \end{pmatrix}
\tag{4.14}
$$

where a, $b \in \{1, 2, 3\}$ and $N \equiv \{N^a\}$. The 4D line-element can then be read off from the above expression,

$$
ds^2 = g_{\mu\nu} dx^\mu dx^\nu = (-N(t)^2 + N^a N_a) dt^2 + 2\, N^a dt\, dx_a + g_{ab} dx^a dx^b
\tag{4.15}
$$

where again a, $b \subset \{1, 2, 3\}$ are spatial indices on Σ_t.

The components g_{ab} of the metric restricted to a leaf of foliation are not the same as the intrinsic metric in a leaf of foliation. The intrinsic metric is related to the projection operator that takes any object $T_{\mu\ldots\nu}$ defined in the full four-dimensional manifold and projects out its component normal to the leaf Σ_t. To understand how to decompose $T_{\mu\ldots\nu}$ into a part $T_\parallel$, which lies only in the hypersurface Σ_t and a part $T_\perp$, orthogonal to Σ_t, we may consider a vector v^μ. The orthogonal component is given by $v_\perp = v^\mu n_\mu$. Similarly the component lying in Σ_t is obtained by projecting the vector along the direction of the shift, so $v_\parallel = v^\mu N_\mu$. Writing a general four-vector as $v^\mu = v_\perp n^\mu + v_\parallel \frac{N^\mu}{|N|}$ (where $|N| = N^\mu N_\mu$ is the norm of the shift vector) and acting on it with $g_{\mu\nu} + n_\mu n_\nu$ we have

$$
(g_{\mu\nu} + n_\mu n_\nu) \left(v_\perp n^\nu + v_\parallel \frac{N^\nu}{|N|} \right) = v_\perp n_\mu (1 + n^\nu n_\nu) + \frac{v_\parallel}{|N|} (N_\mu + n^\nu N_\nu) = v_\parallel \frac{N_\mu}{|N|}.
$$

---

[5] From this expression we can also see that $g_{00} = -N^2 + N^a N_a$ is a measure of the *local* speed of time evolution and hence is a measure of the *local* gravitational energy density.

Since $n^\mu n_\mu = -1$, and $n^\nu N_\nu = 0$ by definition, we are left with only the component of v^μ parallel to Σ_t. We see that $h_{\mu\nu} = g_{\mu\nu} + n_\mu n_\nu$ is the required projection operator. This tensor also happens to correspond to the intrinsic three-metric on Σ_t, induced by its embedding in $\mathcal{M}$,

$$h_{ab} = g_{ab} + n_a n_b,\tag{4.16}$$

where as above $a,\,b \in \{1, 2, 3\}$. The reader might wonder how a rank 3 tensor h_{ab} can be written in terms of a rank 4 object $g_{\mu\nu}$. To understand this, note that the spatial metric can also be written as a rank 4 tensor,

$$h_{\mu\nu} = g_{\mu\nu} + n_\mu n_\nu.$$

However, by construction, the time-time (h_{tt}) and space-time (h_{tx}, h_{ty}, h_{tz}) components vanish and we are left with a rank 3 object. There is no contradiction in writing the spatial metric with either spatial indices ($a, b, \ldots$) or with spacetime indices ($\mu, \nu, \ldots$) as its contraction with another object is non-zero if and only if that object has a purely spatial character.

We have already seen how the Einstein-Hilbert action can be written in terms of the metric $g_{\mu\nu}$ and its derivatives. It makes sense, therefore, that in the case of general relativity, where we have foliated the spacetime into spacelike hypersurfaces, we should take the intrinsic metric on Σ (from now on we drop the t superscript as we will deal with only one, representative, leaf of the foliation) as our configuration or "position" variable. To find the relevant Hamiltonian density we proceed in a manner that parallels the approach in classical mechanics or field theory—namely we perform a Legendre transform to obtain the Hamiltonian function from the Lagrangian. In the case of classical mechanics, given a Lagrangian L dependent on some coordinates q, we see that

$$H[p,\,q] = p\dot{q} - L[q,\,\dot{q}] \quad \text{where} \quad p = \frac{\partial L}{\partial \dot{q}},\tag{4.17}$$

where p is the generalised momentum conjugate to q. Similarly, in the case of scalar field theory, we find that

$$H[\pi,\,\phi] = \int d^4x\,\pi\dot{\phi} - L[\phi,\,\dot{\phi}].\tag{4.18}$$

In the case of GR we find that

$$H[\pi^{\mu\nu},\,h_{\mu\nu}] = \int d^3x\,\pi^{ab}\dot{h}_{ab} - L[h_{ab},\,\dot{h}_{ab}].\tag{4.19}$$

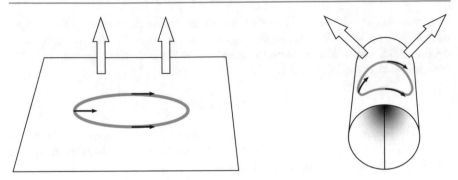

Fig. 4.3 Intrinsic curvature measured by parallel transport (*left*), and extrinsic curvature measured by changes in the normal vectors (*right*). The cylinder in this example has no intrinsic curvature, the same as a flat sheet, since the solid vectors carried around the closed loop are unchanged, but the normal vectors of the cylinder are not all parallel, indicating non-zero extrinsic curvature

In addition to the intrinsic metric h_{ab}, the hypersurfaces Σ also have a tensor which describes their embedding in $\mathcal{M}$, as shown in Fig. 4.3.[6] This object is known as the extrinsic curvature, and is measured by taking the spatial projection of the gradient of the normal vectors to the hypersurface,

$$k_{ab} = h_a{}^c h_b{}^d \nabla_c n_d \equiv D_a n_b \qquad (4.20)$$

where D_a is now the covariant derivative operator which acts only on purely spatial objects. This spatial covariant derivative operator is explored in more detail in Appendix D, Sect. D.1.

As is true in the case of the intrinsic metric, contracting the extrinsic curvature with any object with a time-like component gives zero, $k_{\mu\nu}n^\mu = 0$, implying that the extrinsic curvature is a quantity with only spacelike indices, k_{ab}. Moreover $k_{ab} = k_{(ab)}$ is a symmetric object by virtue of its construction (Sect. D.2).

Due to the properties of the Lie derivative and the purely spatial character of the extrinsic curvature one can show (see Appendix E) that $k_{ab} = \pounds_n h_{ab}$, i.e. the *extrinsic curvature is the Lie derivative of the intrinsic metric with respect to the unit normal vector field n^a*. Now the Lie derivative $\pounds_{\vec{v}} X$ of an object X with respect to a vector field v^a can be interpreted as the *rate of change of X along the integral curves generated by v^a*. By analogy with the definition of p in Eq. (4.17) we might be tempted to identify the extrinsic curvature with the "momentum variable" conjugate to the "position variable" (namely the intrinsic metric). This is not far off the mark. As we will see the conjugate momentum will, indeed, turn out to be a function of k_{ab}.

The Einstein-Hilbert action can be re-written in terms of quantities defined on the spatial hypersurfaces, by making two substitutions. Firstly, and analogously to g, we write h for the determinant of h^{ab} and recognise that the four-dimensional volume

---

[6] The notation $^3\Sigma$ is sometimes used to denote that these are three-dimensional hypersurfaces, however this is redundant in our present discussion.

form $\sqrt{-g}$ is equal to $N\sqrt{h}$ (that is, the three-dimensional volume form multiplied by the distance between hypersurfaces). Secondly, using the Gauss-Codazzi equation,[7]

$$^{(3)}R^{\mu}{}_{\nu\rho\sigma} = h^{\mu}_{\alpha}h^{\beta}_{\nu}h^{\gamma}_{\rho}h^{\delta}_{\sigma}R^{\alpha}{}_{\beta\gamma\delta} - k_{\nu\sigma}k^{\mu}_{\rho} - k_{\nu\rho}k^{\mu}_{\sigma} \tag{4.21}$$

the four-dimensional Ricci curvature scalar R can be re-written in terms of the three-dimensional Ricci scalar $^{(3)}R$ (that is, the Ricci scalar restricted to a hypersurface Σ), and the extrinsic curvature of Σ as

$$R = {}^{(3)}R + k^{ab}k_{ab} - k^2 \tag{4.22}$$

where k is the trace of the extrinsic curvature taken with respect to the 3-metric

$$k := k^{ab}h_{ab} . \tag{4.23}$$

The Gauss-Codazzi relation is a very general result which is true in an arbitrary number of dimensions. The reader with too much time on their hands may wish to derive it for themselves by using the definition of the Ricci scalar in terms of the Christoffel connection and using the 3-metric h^{μ}_{ν} to project quantities in $3+1$ dimensions down to the three dimensions of Σ. By repeating this process with objects living in n and $n+1$ dimensions, one can obtain the version which applies for manifolds of any dimensionality n.

Using these substitutions, the Einstein-Hilbert action can be rewritten in a form that is convenient for identifying the parts which depend only on Σ,

$$S_{\mathrm{EH}} = \int dt\, d^3x\, N\sqrt{h}\left({}^{(3)}R + k^{ab}k_{ab} - k^2\right) = \int dt\, L_{\mathrm{EH}}. \tag{4.24}$$

We next need to find $\dot{h}_{ab}$, which is obtained by taking the Lie derivative (Appendix E) with respect to the vector field t^{μ} which generates time-translations. A detailed derivation is given in Sect. D.3, yielding the result

$$\dot{h}_{ab} = \pounds_{\vec{t}}\, h_{ab} = 2\, N k_{ab} + \pounds_{\vec{N}}\, h_{ab}. \tag{4.25}$$

The conjugate momentum is then found to be

$$\pi^{ab} = \frac{\delta L}{\delta \dot{h}_{ab}} = \sqrt{h}(k^{ab} - k\, h^{ab}). \tag{4.26}$$

---

[7] A derivation of which can be found in Appendix 1.3 of [4].

Substituting these results into Eq. (4.19) we obtain

$$
H[\pi^{ab}, h_{ab}] = \int d^3x\, \pi^{ab} \dot{h}_{ab} - L[h_{ab}, \dot{h}_{ab}]
$$

$$
= \int d^3x\, N \left(-\sqrt{h}^{(3)}R + \frac{1}{\sqrt{h}}(\pi^{ab}\pi_{ab} - \frac{1}{2}\pi^2) \right) - 2N_a D_b \pi^{ab}
$$

$$
= \int d^3x\, N\mathcal{H} - N_a \mathcal{C}^a \tag{4.27}
$$

where for brevity we have adopted the notation

$$
\mathcal{H} = \left(-\sqrt{h}^{(3)}R + \frac{1}{\sqrt{h}}(\pi^{ab}\pi_{ab} - \frac{1}{2}\pi^2) \right) \text{(Hamiltonian constraint)} \tag{4.28a}
$$

$$
\mathcal{C}^a = 2D_b\pi^{ab} \qquad \text{(Diffeomorphism constraint)} \tag{4.28b}
$$

where π is the trace of π^{ab}, and D is the covariant derivative with respect to the 3-metric h_{ab}.

We can reverse the Legendre transform to rewrite the action for GR as

$$
S_{EH} = \int dt L_{EH} = \int dt d^3x \left(\pi^{ab}\dot{h}_{ab} - H[\pi^{ab}, h_{ab}] \right) \tag{4.29a}
$$

$$
= \int dt d^3x \left(\pi^{ab}\dot{h}_{ab} - N\mathcal{H} + N_a\mathcal{C}^a \right). \tag{4.29b}
$$

It is now apparent that the action written in this form is a function of the lapse and shift but *not* their time derivatives. Consequently the Euler-Lagrange equations of motion obtained by varying S_{EH} with respect to the lapse and shift are

$$
\frac{\delta S_{EH}}{\delta N} = -\mathcal{H} = 0\,, \tag{4.30a}
$$

$$
\frac{\delta S_{EH}}{\delta N_a} = \mathcal{C}^a = 0\,, \tag{4.30b}
$$

implying that $\mathcal{H}$ and $\mathcal{C}^a$ are identically zero and are thus to be interpreted as constraints on the phase space! This is nothing more than the usual prescription of Lagrange multipliers—when an action depends only on a configuration variable q but not on the corresponding momentum p, the terms multiplying the configuration variable are constraints on the phase space.

$\mathcal{C}^a$ and $\mathcal{H}$ are referred to as the vector (or diffeomorphism) constraint and the scalar (or Hamiltonian) constraint, respectively. Before we discuss their interpretation, and the reasons for these names, notice that the Hamiltonian density in Eq. (4.27), obtained after performing the $3 + 1$ split of the Einstein-Hilbert action via the ADM procedure [5], is a sum of constraints, i.e. $H_{EH} = N\mathcal{H} - N_a\mathcal{C}^a = 0$. This is a generic feature of diffeomorphism invariant theories.

4.2.1 Physical Interpretation of Constraints

We now briefly describe the form of the Poisson brackets between the various constraints and their physical interpretation. For what follows, it will be helpful to recall some aspects the symplectic formulation of classical mechanics. A system may be described by reference to a phase space in the form of an even-dimensional manifold $\mathcal{K}$ equipped with a symplectic structure (anti-symmetric tensor) $\Omega_{\mu\nu}$. Given any function $f : \mathcal{K} \to \mathbb{R}$ on the phase space, and a derivative operator ∇, there exists a vector field associated with f, given by $X_f^\alpha = \Omega^{\alpha\beta}\nabla_\beta f$. Given two functions f, g on $\mathcal{K}$, the Poisson bracket between the two can be written as

$$\{f, g\} = \Omega^{\alpha\beta}\nabla_\alpha f \nabla_\beta g \tag{4.31}$$

which can also be identified with $-\pounds_{X_f} g = \pounds_{X_g} f$—the Lie derivative (see Appendix E) of g along the vector field generated by f or vice-versa. Thus in this picture, the Poisson bracket between two functions tells us the change in one function when it is Lie-dragged along the vector field generated by the other function (or vice-versa). For more details see [6, Appendix B].

In terms of the intrinsic metric and its conjugate momentum the Poisson bracket between two functions f and g defined on the phase space is given by

$$\{f, g\} = \int d^3x \frac{\delta f}{\delta h_{ab}} \frac{\delta g}{\delta \pi^{ab}} - \frac{\delta f}{\delta \pi^{ab}} \frac{\delta g}{\delta h_{ab}}. \tag{4.32}$$

Since h_{ab} and π^{ab} are fields defined over the three-dimensional manifold Σ, it is necessary to integrate over Σ to obtain a number. Since the diffeomorphism constraint $\mathcal{C}^a = 2D_b\pi^{ab}$ is a function of momenta only, the Poisson bracket of this constraint with the canonical coordinate is given by

$$\{h_{cd}(x'), \xi_a\mathcal{C}^a(x'')\} = -\int d^3x \frac{\delta h_{cd}(x')}{\delta h_{ef}(x)} \frac{\delta \left[2\xi_a D_b\pi^{ab}(x'')\right]}{\delta \pi^{ef}(x)}$$
$$= -\int d^3x\, 2\,\delta_c^e\,\delta_d^f\,\delta_e^a\,\delta_f^b\,\delta(x-x')\,\delta(x''-x)\,D_b\xi_a(x'')$$
$$= -\delta(x'-x'')\,2\,D_d\xi_c \tag{4.33}$$

where ξ_a is a vector field defined on Σ, which serves to "smear out" the constraint $\mathcal{C}^a$ over the manifold so that we get a function defined over the entire phase space, rather than just being defined at each point of Σ. To go from the first line to the second we have integrated by parts and dropped the term which is a pure divergence. This is justified if the field ξ_a has support only on a compact subset of Σ. The constraint $\mathcal{C}^a$ takes the metric h_{ab} to a neighboring point on the phase space, $h_{ab} \to h_{ab} - 2D_b\xi_a$. Using the properties of the Lie derivative, the second term can also be written as $\pounds_\xi h_{ab} = D_a\xi_b + D_b\xi_a$ implying that $h_{ab} \to h_{ab} - \pounds_\xi h_{ab}$, and that therefore $\xi_a\mathcal{C}^a$ is the generator of spatial diffeomorphisms along the vector field ξ_a on the spatial manifold Σ. This is the reason for calling it the "diffeomorphism constraint" in the first place.

Similarly a much more involved calculation along the lines of the one above yields for the Poisson bracket between a function f on the phase space and the "Hamiltonian constraint" $\mathcal{H}$ [7, Sect. I.1.1]

$$\{N\mathcal{H}, f\} = \pounds_{N\vec{n}} f \tag{4.34}$$

i.e. $\mathcal{H}$ generates diffeomorphisms along the vector field $N\vec{n}$ orthogonal to the hypersurface Σ. In other words $\mathcal{H}$ maps functions defined on the hypersurface Σ_t at a given time t to functions on a hypersurface $\Sigma_{t'}$ at a later time t'. This is the reason for referring to $\mathcal{H}$ as the "Hamiltonian constraint"; it generates time evolution of functions on the phase space, the same way the Hamiltonian in classical or quantum mechanics does.

A little later, when we cast GR in the first order formulation we will encounter a third constraint, referred to as the Gauss constraint. We shall discuss the interpretation of the constraints again once the Gauss constraint has been properly introduced, but note here that the Hamiltonian constraint is relevant to the time evolution of the spacelike hypersurfaces, while the other two constraints act spatially (i.e. within the hypersurfaces).

We do not wish to provide more details of the ADM procedure than are strictly necessary. Further details about the ADM splitting and canonical quantization can be found in [2] in the metric formulation, and [5] in the connection formulation.[8]

4.3 Seeking a Path to Canonical Quantum Gravity

In the Hamiltonian formulation one works with a phase space spanned by a set of generalized coordinates q_i, and a set of generalized momenta p_i. For the case of general relativity, the generalised coordinate is the intrinsic metric h_{ab} of the spatial 3-manifold Σ and the extrinsic curvature k_{ab} induced by its embedding in $\mathcal{M}$ determines the corresponding generalized momentum, as per (4.26). For comparison the phase spaces of various classical systems are listed in the following table (Table 4.1)

Table 4.1 Examples of generalised coordinates and momenta for various physical systems

System	Coordinate	Momentum
Simple Harmonic Oscillator	x	p
Ideal Rotor	θ	L_θ
Scalar Field	$\phi(x, t)$	$\pi(x, t)$
Geometrodynamics	h_{ab}	$\pi^{ab} = \sqrt{h}(k^{ab}k_{ab} - k^2)$
Connection Dynamics	$A_a{}^i$	$E^a{}_i$

---

[8] The terms "metric formulation" and "connection formulation" will be defined in Sect. 4.3.1.

Now, given our phase space co-ordinatized by $\{h_{ab}, \pi^{ab}\}$ and the explicit form of the Hamiltonian of GR in terms of the Hamiltonian and diffeomorphism constraints, Eqs. (4.28a) and (4.28b), we may expect that we can proceed directly to quantization by promoting the Poisson brackets on the classical phase space to commutation relations between the operators acting on a Hilbert space H_{GR}:

$$h_{ab} \rightarrow \hat{h}_{ab}, \tag{4.35a}$$

$$\pi^{ab} \rightarrow i\hbar \frac{\delta}{\delta h_{ab}}, \tag{4.35b}$$

$$\left\{ h_{ab}(x), \pi^{a'b'}(x') \right\} = \delta(x - x')\delta_a^{a'}\delta_b^{b'} \rightarrow \left[\hat{h}_{ab}, i\hbar \frac{\delta}{\delta h_{a'b'}} \right] = i\hbar \delta^{a'}{}_a \delta^{b'}{}_b, \tag{4.35c}$$

$$f[h_{ab}] \rightarrow |\Psi_{h_{ab}}\rangle. \tag{4.35d}$$

It should then remain to write the constraints $\mathcal{H}$ and $\mathcal{C}^\mu$ in operator form

$$\mathcal{H}, \mathcal{C}^a \rightarrow \hat{\mathcal{H}}, \hat{\mathcal{C}}^a \tag{4.36}$$

which act upon states $|\Psi_q\rangle$ which would then be identified with the *physical* states of quantum gravity. The physical Hilbert space is a subset of the kinematic Hilbert space which consists of all functionals of the 3-metrics, $|\Psi_{q'}\rangle \in H_{phys} \subset H_{kin}$.

Unfortunately the above prescription is only formal in nature and we run into severe difficulties when we try to implement this recipe. The primary obstacle is the fact that the Hamiltonian constraint stated in Eq. (4.28a) has a *non-polynomial* dependence on the 3-metric via the Ricci curvature $^{(3)}R$. We can see this schematically by noting that $^{(3)}R$ is a function of the Christoffel connection Γ which in turn is a complicated function of h_{ab}:

$$^{(3)}R \sim (\partial \Gamma)^2 + (\Gamma)^2; \quad \Gamma \sim q \partial q \Rightarrow \partial \Gamma \sim \partial q \partial q + q \partial^2 q. \tag{4.37}$$

This complicated form of the constraints raises questions about operator ordering and is also very non-trivial to quantize. Therefore, in this form, the constraints of general relativity are not amenable to quantization.

This is in contrast to the situation with the Maxwell and Yang-Mills fields, which being gauge fields can be quantized in terms of holonomies (see Sect. 3.3), which form a complete set of gauge-invariant variables. An optimist might believe that were we able to rewrite general relativity as a theory of a gauge field, we could make considerably more progress towards quantization than in the metric formulation. This does turn out to be the case as we see in the following sections.

4.3.1 Connection Formulation

Our ultimate goal is to cast general relativity in the mould of gauge field theories such as Maxwell or Yang-Mills. The parallel between covariant derivatives and connections in GR and QFT suggests that gravity may be treated as a gauge field theory with $\Gamma^\rho_{\ \mu\nu}$ as the gauge connection. However, though the Christoffel connection is an affine connection it does not transform as a tensor under arbitrary coordinate transformations (c.f. [2, Chap. 4]) and thus cannot play the role of a gauge connection which should be a covariant quantity.

$\Gamma^\rho_{\ \mu\nu}$ allows us to parallel transport vectors v^μ and,[9] in general, arbitrary tensors (vectors are of course a special case of tensors) i.e. it allows us to map the tangent space T_p at point p to the tangent space $T_{p'}$ at the point p'. The map depends on the path connecting p and p' and it is this fact that allows us to measure *local* geometric properties of a manifold. However, in order to allow the parallel transport of spinors the Christoffel connection is not sufficient.

The Christoffel connection does not "know" about spinor fields of the form $\psi_\mu^{\ I}$ (where I is a Lie algebra index). A theory of quantum gravity which does not know about fermions would not be very useful. Thus we need an alternative to the Christoffel connection which has both these properties; covariance with respect to coordinate transformations, and coupling with spinors.

Up until now we have worked with GR in *second-order* form, i.e. with the metric $g_{\mu\nu}$ as the only configuration variable (hence this is also called the metric formulation). The Christoffel connection $\Gamma^\rho_{\ \mu\nu}$ is determined by the metric compatibility condition,

$$\nabla g_{\mu\nu} = 0. \tag{4.38}$$

The passage to the quantum theory is facilitated by switching to a *first-order* formulation of GR (also called the connection formulation), in which *both* the metric and the connection are treated as independent configuration variables. However due to the problems with the Christoffel connection noted above, we shall choose a first-order formulation in terms of a tetrad or "frame-field" (which we will see shortly takes the role of the metric) and a gauge connection (the "spin connection"), both of which take values in the Lie algebra of the Lorentz group. In the following subsections we will describe the tetrads and the spin connection in some detail, before proceeding to our first example of a first-order formulation of gravity, the Palatini formulation.

The connection formulation exposes a hidden symmetry of geometry as illustrated by the following analogy. The introduction of spinors in quantum mechanics (and the corresponding Dirac equation) allows us to express a scalar field $\phi(x)$ as the "square" of a spinor $\phi = \Psi^i \Psi_i$. In a similar manner the use of the tetrads allows us to write the metric as a square $g_{\mu\nu} = e_\mu^I e_\nu^J \eta_{IJ}$. The transition from the metric to connection variables in GR is analogous to the transition from the Klein-Gordon equation

$$(-\partial_t^2 + \partial^a \partial_a - m^2)\psi = 0 \tag{4.39}$$

to the Dirac equation

$$(i\gamma^\mu \partial_\mu - m)\psi = 0 \tag{4.40}$$

in field theory (where here we have used $c = \hbar = 1$).

---

[9] We will from time to time commit the cardinal sin of conflating a vector with its components.

The connection is a Lie algebra valued one-form $A_\mu{}^{IJ}\tau_{IJ}$ where τ_{IJ} are the generators of the Lorentz group. Our configuration space is then spanned by a tetrad and connection pair, $\{e_\mu^I, A_{IJ}^\mu\}$. The tetrads are naturally identified as mappings between the Lie algebra $\mathfrak{sl}(2, \mathbb{C})$, and the Lie algebra $\mathfrak{so}(3, 1)$ of 4-vectors.

4.3.2   Tetrads

We begin by considering the four dimensional manifold $\mathcal{M}$, introduced in Sect. 4.2, above. As we know, any sufficiently small region of a curved manifold will look flat[10] and so we may define a tangent space to any point P in $\mathcal{M}$. Such a tangent space will be a flat Minkowski spacetime, and the point P may be regarded as part of the worldline of an observer, without loss of generality. This tangent space will be spanned by four vectors, e_μ. Each basis vector will have four components, e_μ^I where $I \in \{0, 1, 2, 3\}$, referred to the locally-defined reference frame (the "laboratory frame" of the observer who's worldline passes through P, with lengths and angles measured using the Minkowski metric).[11] As noted back in Chap. 1, such a set of four basis vectors is referred to as a tetrad or vierbein (German for "four legs").[12] Since the tetrads live in Minkowski space, their dot product is taken using the Minkowski metric. But the dot product of basis vectors is just the metric itself, so the metric of $\mathcal{M}$ at any point is just given by

$$g_{\mu\nu} = e_\mu^I e_\nu^J \eta_{IJ} \tag{4.41}$$

where $\eta_{IJ} = \mathrm{diag}(-1, +1, +1, +1)$ is the Minkowski metric. Taking the determinant of both sides we find that

$$\det(g_{\mu\nu}) = \det(\eta_{IJ})\det(e_\mu^I)^2 = -\det(e_\mu^I)^2 \tag{4.42a}$$

$$\therefore e = \sqrt{-g} \tag{4.42b}$$

where $g \equiv \det(g_{\mu\nu})$ and $e \equiv \det(e_\mu^I)$. Due to this fact the tetrad can be thought of as the "square-root" of the metric.

Tetrads can thus be interpreted as the transformation matrices that map between two sets of coordinates, as can be seen by comparing Eq. (4.41) with the standard form for a coordinate transformation, Eq. (2.16). It is this fact which makes the tetrads a useful tool in modern formulations of GR. Since the components of spinors are defined relative to the flat "laboratory frame" of the tangent space, and tetrads map

---

[10] So long as the manifold is continuous, not discrete. This is an important point to keep in mind for later.

[11] The components can be regarded a internal to the "laboratory frame" tangent space, and hence the choice of indices I, J ... is appropriate.

[12] The similar word *vielbein* ("any legs") is used for the generalisation of this concept to an arbitrary number of dimensions (e.g. triads, pentads).

the metric of this tangent space to the metric of the full four-dimensional spacetime, they serve the role we mentioned above, of allowing us to construct a connection that knows about spinor quantities as well as vectors and tensors. The construction of such a connection will be described below.

As an aside, we note that any vector v^μ can be written as an $\mathfrak{sl}(2, \mathbb{C})$ spinor v_{ij} as

$$v_{ij} := v_\mu e^\mu{}_I \sigma^I{}_{ij} \tag{4.43}$$

where $\sigma^I = \{\mathbf{1}, \sigma^x, \sigma^y, \sigma^z\}$ is a basis of the Lie algebra $\mathfrak{sl}(2, \mathbb{C})$ and i, j are the spinorial matrix indices shown explicitly for clarity.

4.3.3   Choosing a Gauge Group

It is a truth universally acknowledged, that a student in possession of a basic familiarity with loop quantum gravity will be in want of an explanation of the significance of SL(2, $\mathbb{C}$). If we wish to construct a theory that encompasses GR under the framework of gauge field theories we should anticipate that the local symmetries of spacetime will define the gauge group of our quantum gravity theory. To give one example, BF theory (discussed in Sect. 7.2) in three dimensions contains (2+1) general relativity, where choosing SO(3) results in a Riemannian metric,[13] while choosing SO(2,1) results in a Lorentzian one.

We have seen already that gauge field theories are constructed by promoting global gauge symmetries to local symmetries, giving rise to gauge fields which manifest as connection terms in the covariant derivative. The reader will recall from the start of Chap. 2 that special relativity is formulated in a flat spacetime, and hence any transformation of the coordinates applies globally, while GR describes curved spacetime, with SR applying in any sufficiently small region[14] and mappings between coordinate choices in widely-separated regions accounting for the spacetime curvature. In other words, we expect the symmetries of SR to carry over to GR but to be locally applicable, rather than global, and hence expect the existence of a covariant derivative with a connection term. We will discuss the choice of gauge group of this connection term using both a conceptual argument, and a more precise mathematical formulation.

Both arguments involve recognising that rotations in spacetime can be mapped to SL(2, $\mathbb{C}$) transformations. As noted in Chap. 2 the causal structure of spacetime defines a future light-cone and past light-cone at each event. The past light-cone of

---

[13] A Riemannian metric by definition always assigns lengths greater than zero to distinct points in a manifold. This is in contrast to the Minkowski metric, which as per footnote 2 of Chap. 2, may not.

[14] In constructing a theory of quantum gravity we may find that this is not strictly true, as spacetime may not be viably treated as continuous at all length scales, and in fact the concept of a background spacetime may not be valid at all. But these are subtleties to dwell upon in the latter parts of this book. For now, we'll focus on classical theories of spacetime structure.

an observer at any given value of time is the celestial sphere at a fixed distance from the observer. The celestial sphere can be parametrised by the angles θ, ϕ, and any point on a sphere can be stereographically projected onto a plane. For our purposes, this shall be taken to be the complex plane, so that any point on the celestial sphere corresponds with a complex number $\zeta = X + iY$. We can write this as the ratio of two complex numbers $\zeta = \chi / v$. This may seem like an odd thing to do, exchanging one complex number for two, but if we write χ and v as the components of a 2-vector it allows us to let ζ become infinite or zero (corresponding to stereographic projection of the "north pole" and "south pole" of the celestial sphere) by acting on this 2-vector with a transformation that has a finite determinant (and hence keeps the magnitudes of both χ and v within a finite range). Such a linear transformation can be written in the form of a 2×2 matrix with complex components. We can of course write χ and v as functions of θ and ϕ, and vice-versa. So a change of the complex coordinates is equivalent to a coordinate transformation of the real angles θ, ϕ. If we take the determinant of this transformation matrix to be +1 (which we can do, without loss of generality) this is an SL(2, $\mathbb{C}$) transformation. Thus SL(2, $\mathbb{C}$) is the local gauge group of special relativity.

To state this result more precisely, we recognise that the Lorentz group is generated by three rotations, denoted J_a, and three boosts—which we may think of as rotations in the planes defined by the time direction and a spatial direction—denoted K_a. These may be explicitly constructed from the Pauli matrices using the definitions

$$J_a = \frac{1}{2}\sigma_a, \quad K_a = \frac{i}{2}\sigma_a. \tag{4.44}$$

Since the Pauli matrices are traceless 2×2 matrices, the J_a and K_a will generate (via exponentiation, as usual) a group of 2×2 matrices with complex entries, having determinant +1. The group SL(2, $\mathbb{C}$) consists of such matrices, and has six independent parameters, which we can identify with the magnitudes of the rotations and boosts.

The J_a and K_a satisfy the commutation relations

$$[J_a, J_b] = i\epsilon_{abc}J_c, \tag{4.45}$$

$$[K_a, K_b] = -i\epsilon_{abc}J_c, \tag{4.46}$$

$$[J_a, K_b] = i\epsilon_{abc}K_c. \tag{4.47}$$

These are simply the commutation relations of the Lorentz Lie algebra. Thus the correspondence between SL(2, $\mathbb{C}$) and the Lorentz group is established.

The reader should recognise from Eqs. (4.45) and (4.46), that while the commutators of the rotations are linear combinations of rotations, so are the commutators of the boosts. In other words, the boosts are not closed under commutation. However we can define new operators

$$N_a^{\pm} = \frac{J_a \pm iK_a}{2} \tag{4.48}$$

and find that

$$\left[N_a^+, N_b^+\right] = i\epsilon_{abc}N_c^+ , \quad \left[N_a^-, N_b^-\right] = i\epsilon_{abc}N_c^- , \quad \left[N_a^+, N_b^-\right] = 0, \qquad (4.49)$$

hence both the N^+ and N^- independently act like generators of SU(2), and we conclude that the Lie algebra of the Lorentz group is equivalent to two copies of the Lie algebra $\mathfrak{su}(2)$. This leads in a well-known manner to classifying representations (Sect. A.1) of the Lorentz group by pairs of eigenvalues. With J_a and K_a chosen as per Eq. (4.44) we find that $N_a^+ = 0$ and $N_a^- = \frac{1}{2}\sigma_a$, and denote this the $(0, 1/2)$ representation which acts on right-handed spinors. If we had chosen $J_a = \frac{1}{2}\sigma_a$ and $K_a = -\frac{i}{2}\sigma_a$ we would have instead obtained $N_a^+ = \frac{1}{2}\sigma_a$ and $N_a^- = 0$ yielding the $(1/2, 0)$ representation, which acts on left-handed spinors. Note that having one or both generators equal to zero is perfectly acceptable, as exponentiation of zero yields 1, the sole element of the trivial group. Indeed the simplest representation of the Lorentz group is the trivial representation, with all group elements equal to 1. It acts upon Lorentz scalars (since these don't change under a Lorentz transformation i.e. are multiplied by 1). The eigenvalues of $N^+ = 0$ and $N^- = 0$ are both zero, so we refer to this as the $(0, 0)$ representation. Naturally the commutation relations are fulfilled since $[0, 0] = 0$.

But let us return to the matter at hand. Just as a pair of real scalars may be identified with a complex scalar, we may identify $\mathfrak{sl}(2, \mathbb{C})$ as the complexification of $\mathfrak{su}(2)$. Without getting bogged down in details, the take-home message is that SL(2, $\mathbb{C}$) is found to be the universal covering group of the Lorentz group.[15] This should sound familiar to physicists—the well-known fact that SU(2) is a double-cover of SO(3) is another example of such a relationship.

With the correspondence between the Lorentz group and SL(2, $\mathbb{C}$) established we can go on to think about what this means for transformations of spinors. The Lorentz group is a subgroup of the Poincaré group, which consists of translations in addition to the rotations and boosts we have just considered. Dynamics on a flat spacetime can be described by the Poincaré group, however in a general curved spacetime such as we would expect in GR, translational symmetry is broken and only local Lorentz invariance remains as an unbroken symmetry. As discussed in Sect. 2.1 the mapping between local coordinate bases is encoded in the connection. However the Christoffel connection does not allow for the parallel transport of spinors. It is therefore not suitable to be used in constructing a theory of quantum gravity. The simplest candidate that allows for parallel transport of spinors is an $\mathfrak{sl}(2, \mathbb{C})$ valued connection $A_\mu{}^{IJ}$. Such a choice of connection is a logical candidate for casting GR as a gauge theory, and will be referred to as a *spin connection*.

---

[15] It should be borne in mind that the isomorphism is between the algebras, not the groups. The group SU(2), with its relationship to rotations, is compact. This reflects the fact that rotating an object through a finite number of finite rotations can return it to its starting orientation. However the Lorentz group is non-compact, reflecting the fact that even an arbitrarily-large number of finite boosts cannot accelerate an object to the speed of light, and so boosts may be parametrised by a "rapidity" which takes values between negative and positive infinity.

4.3.4 Spin Connection

In order to be able to parallel transport objects with spinorial indices we need a suitable extension of the notion of a covariant derivative which acts on vectors to one which acts on spinors (we follow [8, Appendix B]). The condition for parallel transport of a vector is that its covariant derivative with respect to the Christoffel connection should vanish, i.e.

$$\nabla_k v^i = \partial_k v^i + v^j \Gamma^i_{jk} = 0. \tag{4.50}$$

Similarly the condition for parallel transport of a spinor requires that its covariant derivative with respect to the gauge connection should vanish

$$D_\mu \psi = \partial_\mu \psi + \mathbf{i}g A_\mu \psi = 0 \tag{4.51}$$

where $A_\mu \equiv A^I_\mu t^I$ is the gauge connection. Analogously, given the tetrad e^I_μ and the Christoffel connection $\Gamma^\gamma_{\alpha\beta}$ we define an $\mathfrak{sl}(2, \mathbb{C})$ valued *spin connection* ω^{IJ}_α and use these to construct the *generalised* derivative operator on $\mathcal{M}$ which annihilates the tetrad

$$\mathcal{D}_\alpha e^I_\beta = \partial_\alpha e^I_\beta - \Gamma^\gamma_{\alpha\beta} e^I_\gamma + \omega^I_{\alpha J} e^J_\beta = 0. \tag{4.52}$$

The term "spin connection" may cause some confusion, by tricking newcomers into thinking they have to learn a new concept, when it fact this is nothing more than the notion of parallel transport of a particle along a Wilson line.

Now one would expect that this derivative operator should also annihilate the (internal) Minkowski metric $\eta_{IJ} = e_{\alpha I} e^\alpha_J$ and the spacetime metric $g_{\mu\nu} = e^I_\mu e^J_\nu \eta_{IJ}$. One can check that requiring this to be the case yields that the spin-connection is anti-symmetric $\omega^{\{IJ\}}_\alpha = 0$ and the Christoffel connection is symmetric $\Gamma^\alpha_{[\beta\gamma]} = 0$.

We can solve for $\Gamma^\alpha_{\beta\gamma}$ in the usual manner (see e.g. [2]) to obtain

$$\Gamma^\gamma_{\alpha\beta} = \frac{1}{2} g^{\gamma\delta} \left(\partial_\alpha g_{\delta\beta} + \partial_\beta g_{\delta\alpha} - \partial_\delta g_{\alpha\beta} \right). \tag{4.53}$$

Inserting the above into Eq. (4.52) we can solve for ω to obtain

$$\omega^{IJ}_\alpha = \frac{1}{2} e^{\delta[I} \left(\partial_{[\alpha} e^{J]}_{\delta]} + e^{|\beta|J]} e^K_\alpha \partial_\beta e_{\delta K} \right) \tag{4.54}$$

where the notation on the superscripts indicates that we anti-symmetrize on I and J but not the dummy variable β. Note that in the above expression the Christoffel connection does not occur.

In the definition of $\mathcal{D}$ we have included the Christoffel connection. Ideally, in a gauge theory of gravity, we would not want any dependence on the spacetime connection. That this is the case can be seen by noting that all derivatives that appear in the Lagrangian or in expressions for physical observables are *exterior* derivatives, i.e. of the form $\mathcal{D}_{[\alpha} e^I_{\beta]}$. The anti-symmetrization in the spacetime indices and the symmetry of the Christoffel connection $\Gamma^\gamma{}_{[\alpha\beta]} = 0$ implies that the exterior derivative of the tetrad can be written without any reference to Γ:

$$\mathcal{D}_{[\alpha}e^I_{\beta]} = \partial_{[\alpha}e^I_{\beta]} + \omega_{[\alpha}{}^{IL}e_{\beta]L} = 0. \tag{4.55}$$

We can solve for ω by a trick similar to one used in solving for the Christoffel connection. Following [8, Appendix B], first contract the above expression with $e^\alpha_J e^\beta_K$ to obtain

$$e^\alpha_J e^\beta_K \left(\partial_{[\alpha}e^I_{\beta]} + \omega_{[\alpha}{}^{IL}e_{\beta]L} \right) = 0. \tag{4.56}$$

Now let us define $\Omega_{IJK} = e^\alpha_I e^\beta_J \partial_{[\alpha}e_{\beta]K}$. Performing a cyclic permutation of the indices I, J, K in the above expression, adding the first two terms thus obtained and subtracting the third term we are left with

$$\Omega_{JKI} + \Omega_{IJK} - \Omega_{KIJ} + 2e^\alpha_J \omega_{\alpha IK} = 0. \tag{4.57}$$

This can be solved for ω to yield

$$\omega_{\alpha IJ} = \frac{1}{2}e^K_\alpha [\Omega_{KIJ} + \Omega_{JKI} - \Omega_{IJK}] \tag{4.58}$$

which is equivalent to the previous expression, Eq. (4.54), for ω.

Next we consider the curvature tensors for the Christoffel and the spin connections and show the fundamental identity that allows us to write the Einstein-Hilbert action solely in terms of the tetrad and the spin-connection. The Riemann tensor for the spacetime and the spin connections respectively are defined as

$$\mathcal{D}_{[\alpha}\mathcal{D}_{\beta]}v_\gamma = R_{\alpha\beta\gamma}{}^\delta v_\delta, \qquad \mathcal{D}_{[\alpha}\mathcal{D}_{\beta]}v_I = R_{\alpha\beta I}{}^J v_J. \tag{4.59}$$

Writing $v_\gamma = e^I_\gamma v_I$ and inserting into the first expression we obtain

$$R_{\alpha\beta\gamma}{}^\delta v_\delta = \mathcal{D}_{[\alpha}\mathcal{D}_{\beta]}v_\gamma = \mathcal{D}_{[\alpha}\mathcal{D}_{\beta]}e^I_\gamma v_I = e^I_\gamma R_{\alpha\beta I}{}^J v_J = e^I_\gamma R_{\alpha\beta I}{}^J e^\delta_J v_\delta \tag{4.60}$$

where we have used the fact that $\mathcal{D}_\mu e^I_\nu = 0$. Since the above is true for all v_δ, we obtain

$$R_{\alpha\beta\gamma}{}^\delta = R_{\alpha\beta I}{}^J e^I_\gamma e^\delta_J. \tag{4.61}$$

The Ricci scalar is given by $R = g^{\mu\nu}R_{\mu\nu} = g^{\mu\nu}R_{\mu\delta\nu}{}^\delta$. Using the previous expression we find

$$R_{\mu\delta\nu}{}^\delta = R_{\mu\delta I}{}^J e^I_\nu e^\delta_J. \tag{4.62}$$

Contracting over the remaining two spacetime indices then allows us to write the Ricci scalar in terms of the curvature of the spin-connection and the tetrads,

$$R = R_{\mu\nu}{}^{IJ} e^\mu_I e^\nu_J. \tag{4.63}$$

4.3.5 Palatini Action

The Einstein-Hilbert action, from the discussion in Sect. 4.1, can be written in the form

$$S_{\text{EH}} = \frac{1}{\kappa} \int d^4x \sqrt{-g} g^{\mu\nu} R_{\mu\nu} . \tag{4.64}$$

The Palatini approach to GR starts with this action and treats the metric and the connection as independent dynamical variables. Variation of the action with respect to the metric yields the vacuum field equations $R_{\mu\nu} = 0$, while variation with respect to the connection implies that the connection is the Christoffel connection. Discussion of the Palatini approach in terms of the metric and Christoffel connection can be found in many textbooks (see e.g. [2, Appendix E]).

Having gone to the effort of defining tetrads and the spin connection we now wish to write the action for GR in terms of these variables. We saw in Sect. 4.1 that requirements of covariance and simplicity dictated the form of the action for GR. Similarly our construction of an action based on tetrads and the spin connection is guided by physical considerations. Firstly we want the action to be diffeomorphism invariant. We also require the Lagrangian density to be a four-form, which we can integrate over a four-dimensional spacetime to give a scalar (thus this action is valid only in four dimensions). The curvature of the connection is already a two-form, so (suppressing spacetime indices for simplicity) we include $e^I \wedge e^J \equiv e_{[\mu}{}^I e_{\nu]}{}^J$, which is a two-form.[16] This yields the Palatini action, the simplest diffeomorphism-invariant action one can construct using tetrads and the curvature of the gauge connection. We emphasise that this is not simply S_{EH} rewritten with a change of variables, but a parallel construction. The discussion above is intended to describe the physical intuition behind this construction. It is conventional to use the notation $F_{\mu\nu}^{IJ}$ for the curvature of the spin connection, to yield

$$S_{\text{P}} [e, \omega] = \frac{1}{2\kappa} \int d^4x \star (e^I \wedge e^J) \wedge F^{KL} \epsilon_{IJKL}$$

$$= \frac{1}{4\kappa} \int d^4x \, \epsilon^{\mu\nu\alpha\beta} \epsilon_{IJKL} \, e_\mu{}^I e_\nu{}^J F_{\alpha\beta}{}^{KL} , \tag{4.65}$$

where

$$F^{KL}{}_{\gamma\delta} = \partial_{[\gamma}\omega_{\delta]}{}^{KL} + \frac{1}{2}\left[\omega_\gamma{}^{KM}, \omega_{\delta M}{}^L \right] . \tag{4.66}$$

The similarity between Eqs. (4.64) and (4.65) should be clear, especially when we remember that $g_{\mu\nu} = e_\mu^I e_\nu^J \eta_{IJ}$ (Eq. (4.41)). We also note that this is essentially the action obtained from the BF Lagrangian, Eq. (7.4), with $(e \wedge e)$ replacing E. At this point $F_{\mu\nu}{}^{IJ}$ is the curvature of ω, but it remains to be shown that it satisfies

---

[16] If we use two copies of the curvature tensor then we get Yang-Mills theory $(F \wedge F)$. But that doesn't include the tetrad.

the identity of Eq. (4.63). The equations of motion obtained by varying the Palatini action are

$$\frac{\delta S_{\mathrm{P}}}{\delta \omega_\nu{}^{IJ}} = \epsilon^{\mu\nu\alpha\beta} \epsilon_{IJKL} \, D_\nu \left(e_\alpha{}^I e_\beta{}^J \right) = 0 \,, \tag{4.67a}$$

$$\frac{\delta S}{\delta e_I{}^\mu} = \epsilon^{\mu\nu\alpha\beta} \epsilon_{IJKL} \, e_\nu{}^J F_{\alpha\beta}{}^{KL} = 0 \,. \tag{4.67b}$$

One can see that Eq. (4.67a) is equivalent to the statement that

$$\frac{\delta S[g, \Gamma]}{\delta \Gamma} = 0 \Rightarrow \nabla g = 0 \tag{4.68}$$

therefore in this approach the metric compatibility condition Eq. (4.38) arises as the equation of motion obtained by varying the action with respect to the connection.

Our derivation of Eq. (4.67a) utilized that $F[\omega + \delta\omega] = F[\omega] + \mathcal{D}_{[\omega]}(\delta\omega)$, where $\mathcal{D}_{[\omega]}$ is the covariant derivative defined with respect to the unperturbed connection ω as in Eq. (4.55). The resulting equation of motion, Eq. (4.67a), is then the *torsion-free* or *metric-compatibility* condition which tells us that the tetrad is parallel transported by the connection ω. This then implies that Eq. (4.63) holds, i.e. $F_{\mu\nu}{}^{IJ} \equiv R_{\mu\nu}{}^{IJ}$. The second equation of motion can be obtained by inspection, since F does not depend on the tetrad. Already we see dramatic technical simplification compared to when we had to vary the Einstein-Hilbert action with respect to the metric as in Eq. (4.2).

We will digress at this point, much as we did in Sect. 4.1, in order to show that Eq. (4.67b) is equivalent to Einstein's vacuum equations. We first note that the volume form can be written as

$$\epsilon_{\mu\nu\alpha\beta} = \frac{1}{4!} \epsilon_{PQRS} \, e_{[\mu}{}^P e_\nu{}^Q e_\alpha{}^R e_{\beta]}{}^S. \tag{4.69}$$

Contracting both sides with $e^\nu{}_J$ we find that

$$\epsilon_{\mu\nu\alpha\beta} \, e^\nu{}_J = \frac{1}{4!} \epsilon_{PQRS} \, e_{[\mu}{}^P e_\nu{}^Q e_\alpha{}^R e_{\beta]}{}^S e^\nu{}_J$$

$$= -\frac{1}{3!} \epsilon_{JPQR} \, e_{[\mu}{}^P e_\alpha{}^Q e_{\beta]}{}^R \tag{4.70}$$

where in the second line we have switched some dummy indices and relabelled others. Inserting the right hand side of the above in Eq. (4.67b) and using the fact that Eq. (4.63) implies $F_{\mu\nu}{}^{IJ} \equiv R_{\mu\nu}{}^{IJ}$, we find that

$$\frac{\delta S}{\delta e_I{}^\mu} = \epsilon^{\mu\nu\alpha\beta} \, e_\nu{}^J \epsilon_{IJKL} \, R_{\alpha\beta}{}^{KL}$$

$$= -\frac{1}{3!} \epsilon^{JPQR} \epsilon_{IJKL} \, e_P^{[\mu} e_Q^\alpha e_R^{\beta]} \, R_{\alpha\beta}{}^{KL}$$

$$= \delta_{[I}^P \delta_K^Q \delta_{L]}^R \, e_P^\mu e_Q^\alpha e_R^\beta \, R_{\alpha\beta}{}^{KL}$$

$$= e^{\mu}_{[I} e^{\alpha}_{K} e^{\beta}_{L]} R_{\alpha\beta}{}^{KL}$$

$$= \left(e^{\mu}_{I} e^{\alpha}_{K} e^{\beta}_{L} + e^{\mu}_{K} e^{\alpha}_{L} e^{\beta}_{I} + e^{\mu}_{L} e^{\alpha}_{I} e^{\beta}_{K} \right) R_{\alpha\beta}{}^{KL}$$

$$= e^{\mu}_{I} R + e^{\beta}_{I} R_{\alpha\beta}{}^{\mu\alpha} + e^{\alpha}_{I} R_{\alpha\beta}{}^{\beta\mu}$$

$$= e^{\mu}_{I} R - 2 e^{\beta}_{I} R_{\beta}{}^{\mu} = 0. \tag{4.71}$$

In the first step we have used the result in Eq. (4.70). In the second step we have used the fact that the contraction of two ϵ tensors can be written in terms of anti-symmetrized products of Kronecker deltas. In the third and fourth steps we have simply contracted some indices using the Kronecker deltas and expanded the anti-symmetrized product explicitly. In the fifth and sixth steps we have made use of Eq. (4.61) and the definition of the Ricci tensor as the trace of the Riemann tensor: $R_{\beta}{}^{\mu} = R_{\alpha\beta}{}^{\alpha\mu}$. Contracting the last line of the above with $e^{\nu I}$ and using the fact that $g_{\mu\nu} = e^{I}_{\mu} e^{J}_{\nu} \eta_{IJ}$ we find

$$R_{\mu\nu} - \frac{1}{2} g_{\mu\nu} R = 0. \tag{4.72}$$

Thus the tetradic action in the first-order formulation—where the connection and tetrad are independent variables—is completely equivalent to classical general relativity.

4.3.6   Palatini Hamiltonian and Constraints

Up to this point we have been discussing classical approaches to GR. The Palatini and ADM approaches reproduce Einstein's original formulation of GR, but as mentioned in Sect. 4.3, one would hope that they provide a formulation amenable to canonical quantisation. We can perform a $3 + 1$ split of the Palatini action, Eq. (4.65) and obtain a Hamiltonian which, once again, is a sum of constraints. However, while the resulting formulation appears simpler than that in terms of the metric variables, there are some second class constraints which when solved [8, Sect. 2.4] yield the same set of constraints as obtained in the ADM framework. Thus, the Palatini approach does not appear to yield any substantial improvements over the ADM version as far as canonical quantization is concerned. To proceed to a quantum theory, we must transition to a description of gravity in terms of the Ashtekar variables. But first, let us briefly review the ADM splitting in the tetrad formalism. For this purpose there are two approaches.

The first approach involves repeating the steps in Sect. 4.2, but this time with the first order action (4.65) (where the dynamical variables are the tetrad and the connection), rather than with the Einstein-Hilbert action (4.64). This method is quite tedious and is summarized in Appendix F. Here we present a more direct approach due to Thiemann [7, Sect. I.1.3].

Thiemann's approach is quite simple. It involves starting with the ADM constraints in the metric formulation and rewriting functions of the 3-metric h_{ab} and 3-momentum π_{ab} in terms of the tetrad e^{j}_{a} and the extrinsic curvature one-form

$K^i_a = K_{ab}e^i_b$ (these symbols should not be confused with the Lorentz group boost generators K_a encountered in Sect. 4.3.3.) We already know that the relation between the tetrad one-form and the metric is given by (4.41),

$$h_{ab} = e^i_a e^j_b \delta_{ij}$$

keeping in mind that in $(3 + 1)$ dimensions the correct tensor on the right-hand side would be the Minkowski tensor η_{IJ} rather than the Kroneckar delta. Now under local SO(3) rotations given by the matrix O^i_j, the tetrad changes, $e^i_a \to O^i_j e^j_a$, but the 3-metric h_{ab} remains invariant. Thus, in the tetrad formalism, the action of the rotation group introduces three new degrees of freedom,[17] which were not present in the metric formulation. These extra degrees of freedom can be eliminated by introducing the extrinsic curvature one-form,

$$K^i_a = K_{ab}e^i_b \,, \tag{4.73}$$

where K_{ab} is the extrinsic curvature (4.20) of the 3-manifold Σ. This equation can be inverted to give

$$K_{ab} = e^i_{(a}K^j_{b)}\delta_{ij} \,. \tag{4.74}$$

Since K_{ab} is symmetric, the following constraint on K^i_a must hold:

$$G_{ab} = K^i_{[a}e^j_{b]}\delta_{ij} \,. \tag{4.75}$$

It is convenient to introduce a quantity $\tilde{E}^a_i$ given by the wedge (anti-symmetric) product (see Sect. 3.2 and Appendix B) of two copies of the tetrad,

$$\tilde{E}^a_i = \frac{1}{2}\epsilon^{abc}\epsilon_{ijk}e^j_b e^k_c \,, \tag{4.76}$$

in terms of which (4.75) can be written as

$$G_{ij} = K_{a[i}\tilde{E}^a_{j]} = 0 \,, \tag{4.77}$$

or equivalently, as

$$G_k := \epsilon_k{}^{ij}K_{ai}\tilde{E}^a_j = 0 \,. \tag{4.78}$$

We can now write the 3-metric and 3-momentum in terms of E^a_j and K^i_a as follows;

$$q_{ab} = \det(\tilde{E}^c_l)\,\tilde{E}^i_a \tilde{E}^j_b \tag{4.79a}$$

$$p^{ab} = \det(\tilde{E}^c_l)\,\tilde{E}^a_k \tilde{E}^d_k K^j_{[d}\delta^b_{c]}\tilde{E}^c_j \,. \tag{4.79b}$$

---

[17] In D dimensions, the rotation group has $D(D-1)/2$ degrees of freedom corresponding to the number of independent elements of an antisymmetric $D \times D$ matrix.

When the "Gauss constraint" (4.77), which we first mentioned in Sect. 4.2.1, is satisfied we find that q_{ab}, p^{ab} reduce to the ADM variables h_{ab}, π^{ab}. Using these definitions we can now rewrite the metric ADM constraints (4.28) as

$$\mathcal{G}_i = \epsilon_{ijk} K_a^j \tilde{E}^{ak} \tag{4.80a}$$

$$\mathcal{C}_a = D_b \left[K_a^j \tilde{E}_j^b - \delta_a^b K_c^j \tilde{E}_j^c \right] \tag{4.80b}$$

$$\mathcal{H} = -\det(q) R + \frac{2}{\sqrt{\det(q)}} K_a^j K_b^l \tilde{E}_j^{[a} \tilde{E}_l^{b]} \tag{4.80c}$$

where the first line is the Gauss constraint, the second the diffeomorphism constraint and the third line is the Hamiltonian constraint.

The physical interpretation of these constraints is identical to that given in Sect. 4.2.1, i.e. the diffeomorphism constraint $\xi^a \mathcal{C}_a$ generates spatial diffeomorphisms along the vector field ξ^a on Σ and the Hamiltonian constraint is the generator of diffeomorphisms along the vector field $N\vec{n}$ normal to Σ which corresponds to time-evolution of physical quantities defined on Σ. The only change is the addition of the Gauss constraint (4.78), which acts as the generator of SO(3) rotations. Given an SO(3)-valued form η^i defined on Σ, $\eta^i \mathcal{C}_i$ generates infinitesimal rotations in the triad e_a^i in the "direction" (in the sense of a direction on the SO(3) group manifold) given by η^i.

These constraints satisfy the following Poisson bracket relations:

$$\{K_a^i(x), K_b^j(y)\} = 0 \tag{4.81a}$$

$$\{\tilde{E}_i^a(x), K_b^j(y)\} = \delta_i^j \delta_b^a \delta^3(x, y) \tag{4.81b}$$

$$\{\tilde{E}_i^a(x), \tilde{E}_j^b(y)\} = 0 \tag{4.81c}$$

showing that $\tilde{E}_i^a$ and K_b^j are canonically conjugate variables.

For further details including the calculations of the Poisson bracket structure of these constraints we refer the reader to [7, Sect. I.1.3] or to any of the other reviews listed in the bibliography.

References

1. M.P. Hobson, G.P. Efstathiou, A.N. Lasenby, *General Relativity: An Introduction for Physicists*. (Cambridge University Press, 2006)
2. R.M. Wald, *General Relativity*. (The University of Chicago Press, 1984). ISBN: 9780226870335. https://doi.org/10.7208/chicago/9780226870373.001.0001
3. R. Loll, J. Ambjorn, J. Jurkiewicz, The universe from scratch. Contemp. Phys. **46**, 103–117 (2006). https://doi.org/10.1080/00107510600603344. (arXiv: hep-th/0509010v3)
4. P. Doná, S. Speziale, Introductory lectures to loop quantum gravity (2010). arXiv: 1007.0402
5. Joseph D. Romano, Geometrodynamics vs. connection dynamics. Gen. Rel. Grav. **25**, 759–854 (1993). https://doi.org/10.1007/BF00758384. (arXiv: gr-qc/9303032)
6. A. Ashtekar, Lectures on non-perturbative canonical gravity. Ed. by A. Ashtekar. (World Scientific, 1991). https://doi.org/10.1142/1321

7. T. Thiemann, Introduction to Modern Canonical Quantum General Relativity (2001). https://doi.org/10.48550/gr-qc/0110034. arXiv: gr-qc/0110034
8. Peter Peldan, Actions for gravity, with generalizations: a review. Class. Quant. Grav. **11**, 1087–1132 (1994). https://doi.org/10.1088/0264-9381/11/5/003. (arXiv: gr-qc/9305011)

First Steps to a Theory of Quantum Gravity

5

As discussed in the previous section, we wish to attempt to canonically quantise GR, which means turning the Hamiltonian, diffeomorphism and Gauss constraints into operators and replacing Poisson brackets with commutation relations. This procedure is easier said than done, however. In a practical sense one must be careful with the ordering of operators, and hence constructing appropriate commutation relations is not as easy as one might at first hope. We shall discuss the way forward in outline, before turning to a more detailed discussion of each step. Firstly we simplify the constraints by adopting a complex-valued form for the connection and tetrad variables. These are the Ashtekar variables. Next one performs a $3 + 1$ decomposition to obtain the Einstein-Hilbert-Ashtekar (EHA) Hamiltonian $\mathcal{H}_{\text{EHA}}$ which turns out to be a sum of constraints. We have already seen that these constraints all equal zero, and so when treated as operators they should act upon a state of quantum spacetime, $|\Psi\rangle$ to yield $\mathcal{H}_{\text{EHA}}|\Psi\rangle = 0$. This condition does not force a particular choice of basis for $|\Psi\rangle$ upon us, but it does admit a choice built from objects we are already familiar with—Wilson loops. These loops are then allowed to intersect, to yield area and volume operators of the spacetime. As a result, the states of quantum spacetime come to be represented by graphs whose edges are labelled by representations of the gauge group (for GR this is $SU(2)$). Throughout, the notion of *background independence*,[1] which is central to general relativity, is considered sacrosanct.

The reader interested in the history behind the canonical quantization program, with further mathematical details, is referred to [1].

---

[1] It is important to mention one aspect of background independence that is *not* implemented, *a priori*, in the LQG framework. This is the question of the topological degrees of freedom of geometry. On general grounds, one would expect any four dimensional theory of quantum gravity to contain non-trivial topological excitations at the quantum level. Classically, these excitations would correspond to defects which would lead to deviations from smoothness of any coarse-grained geometry.

© Springer Nature Switzerland AG 2024
S. Bilson-Thompson, *Loop Quantum Gravity for the Bewildered*,
https://doi.org/10.1007/978-3-031-43452-5_5

5.1   Ashtekar Formulation: "New Variables" for General Relativity

We have already discussed the first-order form of GR above. Now let us turn our attention to Ashtekar's complex-valued version of this formalism. We begin with tetradic GR whose action is written in the Palatini form. This action is equivalent to the usual Einstein-Hilbert action *on-shell*, i.e. for configurations which satisfy Einstein's field equations, as shown in Sect. 4.3.5. For dealing with spinors, a formalism defined in terms of connections and tetrads is more useful than one defined in terms of the metric, as shown above. When we perform the ADM splitting of the Palatini action, we switch from variables defined in the full four-dimensional spacetime to the three-dimensional hypersurfaces Σ_t. Hence the tetrads at each point become "triads", $e^I_\mu \to e^i_a$ where $\mu \to a \in \{1, 2, 3\}$, $I \to i \in \{1, 2, 3\}$, and the spin connection is likewise restricted, to become $\Gamma^i_a = \omega_{ajk}\epsilon^{jki}$. The phase space variables of the Palatini picture (e^i_a, Γ^i_a) are the intrinsic metric of the spacelike manifold Σ and a function of its extrinsic curvature respectively, similarly to the situation we noted in Sect. 4.2. Unfortunately in this case the Hamiltonian constraint (Eq. (4.80c)) is still a complicated non-polynomial function and canonical quantization does not appear to be any easier in this formalism.

Ashtekar made the remarkable observation that the form of the constraints simplifies dramatically[2] if instead of the *real* connection $\omega_\mu{}^{IJ}$ one works with a *complex*, self-/anti-self-dual connection (this means that the connection is equal to ± 1 times the dual connection, which is defined in an analogous manner to the dual field strength of Eq. (3.22)). At the heart of the formulation of general relativity as a gauge theory lies a canonical transformation from the triad and connection to the "new" or Ashtekar variables,

$$\tilde{E}^a_i \to \frac{1}{\mathrm{i}}\tilde{E}^a_i, \quad K^i_a \to A^i_a = \Gamma^i_a - \mathrm{i}K^i_a, \tag{5.1}$$

where A^i_a is the Ashtekar-Barbero connection, $K^i_a = k_{ab}e^{bi}$ with k_{ab} the extrinsic curvature of Σ and $\tilde{E}^a_i$ is the variable introduced previously in (4.76).

Both A^i_a and $\tilde{E}^a_i$ admit SU(2) rotations with respect to the internal indices (and hence the choice of densitised triads is non-unique). We can therefore treat the Ashtekar formulation of gravity as an SU(2) gauge theory. This is consistent with our previous discussion about the choice of gauge group for gravity (Sect. 4.3.3), as SU(2) is a subgroup of SL(2, $\mathbb{C}$).

Given this choice of variables, the constraints simplify to

$$\mathcal{G}_i = D_a\tilde{E}^a_i \quad \text{(Gauss constraint)} \tag{5.2a}$$

$$\mathcal{C}_a = \tilde{E}^b_i F^i{}_{ab} - A^i_a\mathcal{G}_i \qquad \text{(Diffeomorphism constraint)} \tag{5.2b}$$

$$\mathcal{H} = \epsilon^{ij}{}_k \tilde{E}^a_i \tilde{E}^b_j F^k{}_{ab} \quad \text{(Hamiltonian constraint)} \tag{5.2c}$$

---

[2] For the detailed derivation of these constraints starting with the self-dual Lagrangian see e.g. [2, Sect. 6.2].

Comparing these to the form of the Palatini constraints (4.80a), we see that the Gauss constraint now takes the form of a net divergence of the triad "electric" field, in analogy with the form of the Gauss law in electromagnetism. The diffeomorphism constraint is now linear in the triad field. The greatest simplification is seen in the Hamiltonian constraint which is now only quadratic in the triad, whereas previously, due to the presence of the $1/\det(q)$ term, it had a non-polynomial dependence on the triad. This makes quantisation *feasible*. In fact [3], it turns out that the exponential of the Chern-Simons invariant on the manifold is an exact solution of all three constraints! (See Appendix G for more details.)

The phase space configuration and momentum variables are $\tilde{E}_i^a$ and the spatial connection $A^i{}_a$. The second class constraints which were present in the Palatini framework must now vanish due to the Bianchi identity (see [4, Sect. 2.4–2.5]) and the diffeomorphism constraint becomes a polynomial quadratic function of the momentum variables—in this case the triad. We thereby obtain a form for the constraints which is polynomial in the coordinates and momenta and thus amenable to methods of quantization used for quantizing gauge theories such as Yang-Mills. The resulting expression for the Einstein-Hilbert-Ashtekar Hamiltonian of GR is

$$\mathcal{H}_{\text{EHA}} = N^a \mathcal{C}_a + N\mathcal{H} + T^i \mathcal{G}_i = 0 \tag{5.3}$$

where $\mathcal{C}_a$, $\mathcal{H}$ and $\mathcal{G}_i$ are the vector, scalar and Gauss constraints respectively. The terms N_i^a and N are the shift and lapse, while T^i is a lie-algebra valued function over our spatial surface which encodes the freedom we have in choosing the gauge for the gauge connection. As in Sect. 4.2.1 we can calculate the Poisson brackets between these constraints and the canonical variables. Doing so verifies the intuition gained from Sect. 4.2.1. The Poisson brackets of a function f with the Hamiltonian and diffeomorphism constraints gives

$$\{f, \mathcal{H}\} = \pounds_{N\tilde{n}} f, \qquad \{f, \xi^a \mathcal{C}_a\} = \pounds_{\tilde{\xi}} f, \tag{5.4}$$

implying that as expected $\mathcal{H}$ and $\mathcal{C}_a$ generate time-evolution and spatial diffeomorphism respectively. Introducing the gauge degrees of freedom has also led to the introduction of a third constraint $\mathcal{G}_i$, for whose Poisson bracket we have

$$\{f, T^i \mathcal{G}_i\} = -\tilde{E}_i^a D_a T^i, \tag{5.5}$$

implying that $\mathcal{G}_i$ corresponds to the generators of gauge rotations.

It is instructive to compare the above form of the constraints to their metric counterparts in Eq. (4.28) which are reproduced below for the reader's convenience:

$$\mathcal{H} = \left(-\sqrt{h}^{(3)}R + \frac{1}{\sqrt{h}}(\pi^{ab}\pi_{ab} - \frac{1}{2}\pi^2)\right),$$
$$\mathcal{C}^a = 2D_b\pi^{ab}.$$

The price to be paid for this simplification is that the theory we are left with is no longer the theory we started with—general relativity with a manifestly real metric geometry. The connection is now a complex connection. However the new concoction is also not too far from the original theory and can be derived from an action. That this is the case was shown independently by Jacobson and Smolin [5] and by Samuel [6]. They completed the analysis by writing down the Lagrangian from which Ashtekar's form of the constraints would result:

$$S_\pm [e, A] = \frac{1}{4\kappa} \int d^4x \, {}^\pm\Sigma^{\mu\nu}{}_{IJ} {}^\pm F_{\mu\nu}{}^{IJ}. \tag{5.6}$$

Here ${}^\pm F$ is the curvature of a *self-dual* (anti-self-dual) four-dimensional connection ${}^\pm A$ one-form, which we will discuss more in the next subsection. The field ${}^\pm\Sigma$ is the self-dual (anti-self-dual) portion of the two-form $\tilde{e}^I \wedge \tilde{e}^J$. The Palatini action is then simply given by the real part of the self-dual (or anti-self-dual) action,

$$S_\mathrm{P} = \mathbf{Re}[S_\pm]. \tag{5.7}$$

5.2   The Barbero-Immirzi Parameter

In the previous section we saw that the transformation (5.1) from the Palatini variables $\{K_a^i, E_i^a\}$ to the Ashtekar variables $\{A_a^i, E_i^a\}$ is of the form

$$\tilde{E}_i^a \to \frac{1}{\mathbf{i}} \tilde{E}_i^a; \qquad K_a^i \to A_a^i = \Gamma_a^i - \mathbf{i} K_a^i.$$

While this leads to simplification of the constraints, the presence of the unit imaginary $\mathbf{i} = \sqrt{-1}$ in the transformation rule also makes the theory complex! In order to obtain physical results—corresponding to a metric valued in $\mathbb{R}$ instead of in $\mathbb{C}$—we must impose some restrictions on the possible solutions of the theory. If we use the notation $X^\bullet$ to represent the time derivative of X, then solutions must satisfy not only the constraints (5.2a)–(5.2c), but also the so-called "reality conditions",

$$\tilde{E}_i^a \tilde{E}_j^b \delta^{ij} \in \mathbb{R}, \tag{5.8a}$$

$$\left(\tilde{E}_i^a \tilde{E}_j^b \delta^{ij} \right)^\bullet \in \mathbb{R}. \tag{5.8b}$$

The first of these is simply the requirement that the metric constructed from the triad field be real. The second says that the metric should remain real under time evolution.

As first pointed out in [7–9], the Ashtekar variables are a particular case of a more general transformation,[3]

$$E_i^a \to \frac{1}{\gamma} E_i^a, \qquad K_a^i \to A_a^i = \Gamma_a^i - \gamma K_a^i \tag{5.9}$$

---

[3] The transformation to new variables, as implemented by most authors, including Barbero and Immirzi, does not involve changing the triad. In this case the Poisson brackets between the new variables picks up a factor of γ. However if we transform the triad also, as is done here following [1], the factor of γ cancels out when taking the Poisson brackets.

where γ, the so-called "Barbero-Immirzi" parameter (often referred to as just the Immirzi parameter), is an arbitrary complex number with a significant physical interpretation. It is related to the size of a quantum of area, measured in Planck units, and thus to the area operators mentioned in this chapter's introduction [10, Sect. 6.6.2]. For the particular choice of $\gamma = \mathbf{i}$, the above variables reduce to Ashtekar's original form. For any other choice of γ, however, the resulting variables are just as valid because the transformation remains canonical, i.e. the Poisson brackets before and after the transformation continue to remain the same, namely

$$\left\{ K_a^i(x), \tilde{E}_j^b(y) \right\} = \left\{ A_a^i(x), {}^{(\gamma)}\tilde{E}_j^b(y) \right\} = \kappa \delta_j^i \delta_a^b \delta(x, y). \qquad (5.10)$$

The Barbero-Immirzi parameter, and constraining its possible values, has been an area of active discussion in the (loop and allied) quantum gravity community for some time. Specific topics of discussion include determining its value from comparison with the calculation of the Bekenstein-Hawking gravity of a black hole [11–17], four-fermion interactions sourced by non-zero value of γ [18–20], effects of renormalization on values of γ [15,21], possible relationship to the Standard Model [22], its role in obtaining a generalization of the Kodama state (Appendix G) which overcomes difficulties first pointed out by Witten [23–25], its role in determining the strength of topological interactions in LQG of the sort encountered in the Peccei-Quinn mechanism of the Standard Model [26–29], and more recently a possible holographic interpretation of γ [30,31]. This list is not meant to be exhaustive and any errors and emissions of significant contributions related to γ are solely the result of the authors' ignorance.

5.3 To Be or Not to Be (Real)

Before proceeding to the details of the quantization procedure let us address the controversy over which is preferable, real variables or complex ones. The reality of the Ashtekar variables depends on the reality of the Barbero-Immirzi parameter—if γ is complex (or real), then so are the Ashtekar variables. We have chosen to introduce the complex variables for historical reasons, and because the self-dual variables have great pedagogical value for explaining the steps leading to the simplified form of the ADM constraints. However the newcomer to LQG (at whom this book is aimed) should be aware that the choice between real or complex variables is not immediately obvious. Over time conventional wisdom has favored the real variables, primarily because they allow us to construct the kinematical Hilbert space of $SU(2)$ spin networks, and some may question the need to discuss the complex variables in any great detail in an introductory review. But it is worth noting that in recent years self-dual variables have made something of a comeback in works by Wieland, Frodden et al., and Pranzetti. Each choice has its pros and cons.

The advantage of using a real value for γ is that the new variables and the resulting constraints remain real, avoiding the need to impose reality conditions (5.8) on solutions of the constraints. Secondly, a real γ avoids difficulties that arise when moving to the quantum theory—since with $\gamma = \mathbf{i}$, the gauge group of the theory is $SL(2, \mathbb{C})$, which is non-compact and it is not clearly understood how to perform integration over non-compact groups.

The disadvantage of a real value for γ is that the form of the Hamiltonian constraint is no longer as simple as given in (5.2c), and picks up another term quadratic in both the triad and extrinsic curvature,

$$\mathcal{H} = \epsilon^{ij}{}_k \tilde{E}^a_i \tilde{E}^b_j F^k{}_{ab} - 2(1 + \gamma^2) \tilde{E}^{[a}_i \tilde{E}^{b]}_j K^i_a K^j_b \approx 0. \qquad (5.11)$$

In contrast we can see that when $\gamma = \mathbf{i}$, the second term in the above vanishes and we are left with the usual Ashtekar form of the constraint.

A complex value of γ implies that the spectrum of the area operator (to be studied in greater detail in Sect. 6.3) will also contain complex eigenvalues. It is not clear what physical interpretation one can assign to complex areas or complex volumes. Moreover the structure of the kinematical Hilbert space of LQG (to be discussed later in Chap. 6) is understood only for the case of real γ. If γ is taken to be complex then the entire technology of spin networks—using which, for instance, the black hole entropy calculation (Sect. 8.1) is performed—is rendered unusable.

On the other hand, retaining a complex γ means that the spatial connection has an interpretation as a spacetime connection since it transforms correctly under diffeomorphisms [6,32,33], whereas this is not true for the real or "Ashtekar-Barbero" connection. The Hamiltonian constraint is polynomial which is one of the principle motivations and advantages for going from metric dynamics to connection dynamics.

Wieland [34,35] has shown that starting from the complex variables (with $SL(2, \mathbb{C})$ as gauge group) one can perform the canonical quantization procedure and obtain the *same* kinematic Hilbert space as in the SU(2) case. Thus the earlier concerns regarding the viability of complex variables vis-a-vis the existence of the kinematical Hilbert space would appear to have been resolved.

Frodden, Perez, and Ghosh [36] have shown that when the dimension of the Hilbert space of SU(2) Chern-Simons gauge theory, which describes the dynamics of a quantum isolated horizon (QIH) [37,38], is analytically continued to complex values, its asymptotic behavior has an exponential dependence on the horizon area. In this way the Bekenstein-Hawking entropy is recovered in the semiclassical limit.

Pranzetti [39] has demonstrated that in order to provide a *geometrical notion* of the temperature of a QIH one must work with a complex value of the Barbero-Immirzi parameter. Taking $\gamma = \mathbf{i}$ and requiring that the horizon state satisfying the QIH boundary condition be a Kubo-Martin-Schwinger (KMS) state (i.e. a thermal equilibrium state) leads to the formula for the temperature of the horizon. The Boltzmann and von-Neumann entropies can also be calculated and in the semi-classical limit both yield the expression for the Bekenstein-Hawking entropy.

With these observations in hand, it is certainly too soon to consign the complex variables to being merely a historical footnote in the development of LQG.

5.4   Loop Quantization

As noted above, the program of loop quantum gravity involves the following steps:

1. Write GR in connection and tetrad variables (in first order form).
2. Perform a $3 + 1$ decomposition to obtain the Einstein-Hilbert-Ashtekar Hamiltonian $\mathcal{H}_{\mathrm{EHA}}$ which turns out to be a sum of constraints.
3. Obtain a quantized version of the Hamiltonian whose action on elements of the physical space of states yields $\mathcal{H}_{\mathrm{EHA}}|\Psi\rangle = 0$.
4. Identify an appropriate basis for the physical states of spacetime.

The first two steps have been thoroughly covered. So now, after a fairly lengthy digression, we are ready to return to the task mentioned in Sect. 4.3, rewriting the constraints in operator form, and identifying the physical states of quantum gravity. The first part of this process was completed in Eqs. (5.2a)–(5.2c).

The following exposition only gives us a bird's eye view of the process of canonical quantization. The reader interested in the mathematical details of and the history behind the canonical quantization program is referred to [1].

5.5   Canonical Quantization

To find solutions of the equations of motion we want to find states $\Psi[A]$ such that they are acted upon appropriately by the constraints. This means that they satisfy

$$\hat{\mathcal{H}}|\Psi\rangle = 0$$
$$\hat{\mathcal{C}}_a|\Psi\rangle = 0$$
$$\hat{\mathcal{G}}_i|\Psi\rangle = 0$$

The Gauss constraint tells us that $\Psi[A]$ should be gauge-invariant functions of the connection. The diffeomorphism constraint is telling us that $\Psi[A]$ should be invariant under diffeomorphisms of the paths along which the connection lies. These constraints taken together do not impose a particular choice of $\Psi[A]$ upon us, but they do admit Wilson loops as one possible, and particularly convenient, choice.

Let us consider solutions of the form $\Psi[A] = \sum_\lambda \Psi[\lambda]W_\lambda[A]$. A given state will therefore be a sum of loops. These loops may in general be knotted, and hence topologically distinct from each other. Such states will satisfy the Gauss constraint, as Wilson loops are gauge-invariant. They will also satisfy the diffeomorphism constraint. In fact, diffeomorphism invariance actually helps us reduce the number of basis states, thereby avoiding a potentially awkward problem. In a theory with a fixed background and a well-defined metric any tiny change in the shape of a Wilson loop will lead to a different loop holonomy, since parallel transport is path-dependent. If different loops are taken to be the orthonormal basis states, this means that each deformation of a loop results in a new state, orthonormal to every other loop. But in a diffeomorphism-invariant theory it is not possible to distinguish between any two loops that may be smoothly deformed into each other, and hence the space of loops consists of only a single member of each topological equivalence class.

Now we must ask whether Wilson loops satisfy the Hamiltonian constraint. Firstly we observe that the triads (or tetrads when we are working in four dimensions) are the conjugate momenta to the connection. In quantum mechanics the operator for the momentum corresponds to derivation with respect to the position coordinate, $p \to \hat{p} = -i\hbar\frac{\partial}{\partial q}$. Similarly the quantum operator for the triad (or tetrad!) is given by the derivative with respect to the connection, hence $e_a{}^i \to -i\hbar\frac{\partial}{\partial A_a{}^i}$. The action of $\hat{\mathcal{H}}$ on a Wilson loop is therefore

$$\hat{\mathcal{H}}W_\lambda[A] = \epsilon^{ij}{}_k \frac{\delta}{\delta A_a^i} \frac{\delta}{\delta A_b^j} F^k{}_{ab} W_\lambda[A]. \tag{5.12}$$

The exact form of the resulting functional derivatives is not important for the moment. However, as will be discussed below (Eq. (6.10)) the loop holonomies contain a term representing the tangent vector to the curve along which the loop holonomy is evaluated. But the tangent vector to the loop is $\dot{\lambda} = d\lambda/ds$ where s parametrises points along the loop. Due to the exponential form of the loop holonomy the derivatives pull out factors of $\dot{\lambda}$. Then since $F^k{}_{ab} = -F^k{}_{ba}$ it follows that summation over the indices of the curvature yields zero, and hence $F^k{}_{ab}\dot{\lambda}^a\dot{\lambda}^b = 0$, confirming that the Hamiltonian constraint is satisfied.

This loop basis gives us a picture of spacetime at the smallest scale, consisting of closed paths carrying representations of SU(2). It now remains to interpret the loop basis in terms of physical observables.

References

1. T. Thiemann, Introduction to Modern Canonical Quantum General Relativity (2001). https://doi.org/10.48550/gr-qc/0110034. arXiv: gr-qc/0110034
2. Joseph D. Romano, Geometrodynamics vs. connection dynamics. Gen. Rel. Grav. **25**, 759–854 (1993). https://doi.org/10.1007/BF00758384. (arXiv: gr-qc/9303032)
3. Hideo Kodama, Holomorphic wave function of the Universe. Phys. Rev. D **42**, 2548–2565 (1990). https://doi.org/10.1103/PhysRevD.42.2548. (url: dx.doi.org/10.1103/PhysRevD.42.2548)
4. Peter Peldan, Actions for gravity, with generalizations: a review. Class. Quant. Grav. **11**, 1087–1132 (1994). https://doi.org/10.1088/0264-9381/11/5/003. (arXiv: gr-qc/9305011)
5. T. Jacobson, L. Smolin, Covariant action for Ashtekar's form of canonical gravity. Class. Quant. Gravity **5**(4), 583+ (1988). ISSN: 0264-9381. https://doi.org/10.1088/0264-9381/5/4/006
6. J. Samuel, A lagrangian basis for Ashtekar's reformulation of canonical gravity. Pramana **28**(4), L429–L432 (1987). ISSN: 0304-4289. https://doi.org/10.1007/BF02847105
7. Fernando Barbero, Real Ashtekar variables for lorentzian signature space-times. Phys. Rev. D **51**, 5507–5510 (1995). https://doi.org/10.1103/PhysRevD.51.5507. (arXiv: gr-qc/9410014)
8. Fernando Barbero, From Euclidean to Lorentzian general relativity: the real way. Phys. Rev. D **54**, 1492–1499 (1996). https://doi.org/10.1103/PhysRevD.54.1492. (arXiv: gr-qc/9605066)
9. Giorgio Immirzi, Real and complex connections for canonical gravity. Class. Quant. Grav. **14**, L177–L181 (1996). https://doi.org/10.1088/0264-9381/14/10/002. (arXiv: gr-qc/9612030)

10. C. Rovelli, *Quantum Gravity*. (Cambridge University Press, 2004). https://doi.org/10.1017/CBO9780511755804

11. K. Krasnov, On the constant that fixes the area spectrum in canonical quantum gravity. Class. Quant. Grav. **15**, L1–L4 (1998). https://doi.org/10.1088/0264-9381/15/1/001. arXiv: gr-qc/9709058

12. A. Ashtekar et al., Quantum geometry and black hole entropy. Phys. Rev. Lett. **80**, 904–907 (1998). https://doi.org/10.1103/PhysRevLett.80.904. arXiv: gr-qc/9710007

13. O. Dreyer, Quasinormal modes, the area spectrum, and black hole entropy. Phys. Rev. Lett. **90** (2003). https://doi.org/10.1103/PhysRevLett.90.081301. arXiv: gr-qc/0211076

14. M. Domagala, J. Lewandowski, Black hole entropy from Quantum Geometry". Class. Quant. Grav. **21**, 5233–5244 (2004). https://doi.org/10.1088/0264-9381/21/22/014. arXiv: gr-qc/0407051

15. T. Jacobson, Renormalization and black hole entropy in Loop Quantum Gravity. Class. Quant. Grav. **24**, 4875–4879 (2007). https://doi.org/10.1088/0264-9381/24/18/N02. arXiv: 0707.4026

16. A. Majhi, Microcanonical Entropy of Isolated Horizon and the Barbero-Immirzi parameter. Class. Quant. Grav. **31**, 095002 (2014). https://doi.org/10.1088/0264-9381/31/9/095002. arXiv: 1205.3487

17. D. Pranzetti, H. Sahlmann, Horizon entropy with loop quantum gravity methods. Phys. Lett. B **746**, 209–216 (2015). https://doi.org/10.1016/j.physletb.2015.04.070. arXiv: 1412.7435. http://adsabs.harvard.edu/abs/2015PhLB..746..209P

18. A. Perez, C. Rovelli, Physical effects of the Immirzi parameter in loop quantum gravity. Phys. Rev. D **73**(4), 044013 (2006). https://doi.org/10.1103/PhysRevD.73.044013. arXiv: gr-qc/0505081

19. S.H.S. Alexander, D. Vaid, Gravity Induced Chiral Condensate Formation and the Cosmological Constant (2006). https://doi.org/10.48550/arXiv.hep th/0609066. arXiv: hep-th/0609066. http://arXiv.org/abs/hep-th/0609066

20. S. Alexander, D. Vaid, A fine tuning free resolution of the cosmological constant problem (2007). https://doi.org/10.48550/arXiv.hep-th/0702064. arXiv: hep-th/0702064

21. C-H. Chou, R-S. Tung, H-L. Yu, Origin of the Immirzi parameter. Phys. Rev. D **72**, 064016 (2005). https://doi.org/10.1103/PhysRevD.72.064016. arXiv: gr-qc/0509028

22. B. law Broda, M.1 Szanecki, A relation between the Barbero-Immirzi parameter and the standard model. Phys. Lett. B **690**(1), 87–89 (2010). ISSN: 03702693. https://doi.org/10.1016/j.physletb.2010.05.004. arXiv: arXiv:1002.3041

23. A. Randono, Generalizing the Kodama State I: Construction (2006). https://doi.org/10.48550/arXiv.gr-qc/0611073. arXiv: gr-qc/0611073

24. A. Randono, Generalizing the Kodama state II: properties and physical interpretation (2006). https://doi.org/10.48550/arXiv.gr-qc/0611074. arXiv: gr-qc/0611074

25. A. Randono, In Search of Quantum de Sitter Space: Generalizing the Kodama State. Ph.D. thesis. University of Texas at Austin (2007). https://doi.org/10.48550/arXiv.0709.2905

26. S. Mercuri, From the Einstein-Cartan to the Ashtekar-Barbero canonical constraints, passing through the Nieh-Yan functional. Phys. Rev. D **77**(2) (2008). ISSN: 1550-7998. https://doi.org/10.1103/physrevd.77.024036. arXiv: 0708.0037

27. G. Date, R.K. Kaul, S. Sengupta, Topological interpretation of Barbero-Immirzi parameter. Phys. Rev. D **79**(4) (2008). ISSN: 1550-7998. https://doi.org/10.1103/physrevd.79.044008. arXiv: 0811.4496

28. S. Mercuri, Peccei-Quinn mechanism in gravity and the nature of the Barbero-Immirzi parameter. Phys. Rev. Lett. **103**(8) (2009). ISSN: 0031-9007. https://doi.org/10.1103/PhysRevLett.103.081302. arXiv: arXiv:0902.2764

29. J. Magueijo, D.M.T. Benincasa, Chiral vacuum fluctuations in quantum gravity. Phys. Rev. Lett. **106**, 121302 (2011). https://doi.org/10.1103/PhysRevLett.106.121302. arXiv: 1010.3552

30. M. Sadiq, A correction to the Immirizi parameter of SU(2) spin networks. Phys. Lett. B **741**, 280–283 (2015). https://doi.org/10.1016/j.physletb.2015.01.004. (arXiv.org/abs/1409.7726)

31. M. Sadiq, The holographic principle and the Immirzi parameter of loop quantum gravity (2015). https://doi.org/10.48550/arXiv.1510.04243. arXiv: 1510.04243

32. S. Alexandrov, On choice of connection in loop quantum gravity. Phys. Rev. D **65**(2) (2005). ISSN: 0556-2821. https://doi.org/10.1103/physrevd.65.024011. arXiv: gr-qc/0107071
33. J. Samuel, Is Barbero's Hamiltonian formulation a Gauge Theory of Lorentzian Gravity? Class. Quant. Grav. **17**, L141–L148 (2000). https://doi.org/10.1088/0264-9381/17/20/101. arXiv: gr-qc/0005095
34. W. Wieland, Complex Ashtekar variables and reality conditions for Holst's action. Ann. H. Poincaré 1–24 (2010). https://doi.org/10.1007/s00023-011-0134-z. arXiv: 1012.1738
35. W. Wieland, Complex Ashtekar variables, the Kodama state and spinfoam gravity (2011). https://doi.org/10.48550/arXiv.1105.2330. arXiv: 1105.2330
36. E. Frodden et al., Black Hole Entropy from complex Ashtekar variables. Europhys. Lett. **107**, 10005 (2014). https://doi.org/10.1209/0295-5075/107/10005. arXiv: 1212.4060
37. J. Engle et al., Black hole entropy from an SU(2)-invariant formulation of Type I isolated horizons. Phys. Rev. D **82**, 044050 (2010). https://doi.org/10.1103/PhysRevD.82.044050. arXiv: 1006.0634
38. J. Engle, K. Noui, A. Perez, Black hole entropy and SU(2) Chern-Simons theory. Phys. Rev. Lett. **105**, 031302 (2010). https://doi.org/10.1103/PhysRevLett.105.031302. arXiv: 0905.3168
39. D. Pranzetti, Black hole entropy from KMS-states of quantum isolated horizons. Phys. Rev. D **89**, 104046 (2014). https://doi.org/10.1103/PhysRevD.89.104046. arXiv: 1305.6714

Kinematical Hilbert Space

6

In light of the discussion at the end of the previous chapter, we see that spacetime can be represented by a wide range of loop states. These may consist of simple closed loops. The loops may be linked through each other. They may also be knotted, and hence classified by knot invariants. And the loops may intersect, creating vertices at which three or more Wilson lines meet. Historically the importance of all these possibilities has been considered, and continues to be assessed. We will simply take the view that a general loop state can have all of the properties listed above, and turn our attention to understanding the kinematics inherent to this loop picture of spacetime.

Before we dive straight in to the main subject matter of this chapter it is useful to introduce a simple model of space, and examine its kinematics. This will allow us to familiarise the reader with the terminology that will follow, as well as introducing the basic concepts by which areas and volumes of spacetime regions can be defined. This also sets the stage for a discussion of dynamics, to be pursued in Chap. 7, in a low-dimensional case that is easy to visualise.

Throughout this chapter several expressions arise which involve spin operators and components, and the reader should bear in mind that we will be relying heavily on the convention that j, J (i.e. in bold) refer to spins, while j, J are indices. However we will still write "3-j symbol" and "6-j symbol" as these are the *names* of particular mathematical structures, rather than the notation that denotes their values.

6.1 Kinematics via a Toy Model

Consider a two-dimensional manifold, embedded in three dimensions. The topology of the mainfold could be arbitrary but to develop useful terminology we are only concerned with a small region of it, so it may as well not have any handles or holes. Such a manifold is what we would commonly think of in day-to-day life as a surface, and can be approximated by a set of flat triangles. By "flat triangles" we mean that the region within the three edges of each triangle is a subset of $\mathbb{R}^2$. Of course, the triangles

© Springer Nature Switzerland AG 2024
S. Bilson-Thompson, *Loop Quantum Gravity for the Bewildered*,
https://doi.org/10.1007/978-3-031-43452-5_6

Fig. 6.1 An example of a
triangulation of a 2D surface

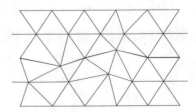

will not necessarily be coplanar with each other, and so the whole arrangement will resemble interlocking triangular scales on the skin of some animal. The surface being approximated may have some (extrinsic and/or intrinsic) curvature, and so we describe the triangular approximation as piecewise flat. The dimensionality of the triangles (being 2) is one less than that of the space in which they are embedded. The triangles themselves meet at 1-dimensional edges, and the edges meet at 0-dimensional vertices. Each triangle is referred to as a simplex (plural, simplices). Therefore the surface is covered by a pattern of simplices clustered around vertices, as depicted in Fig. 6.1. We will refer to this entire arrangement of simplices, edges, and vertices as a triangulation, denoted Δ. The lengths x_1, x_2, x_3 of the edges of each two-dimensional simplex must obey the triangle inequality, $x_1 + x_2 \geq x_3$ (and of course this relation must hold true for any permutations of 1, 2, and 3).

If the surface being triangulated is flat (i.e. has no intrinsic curvature) traversing a closed path[1] will mean turning through a total of 360° or 2π. However if some of the triangular simplices adjacent to the vertex at the centre of such a closed path were removed and the remaining simplices reconnected, a closed path would traverse less than 2π (or more, if instead extra simplices were added). In this case the triangulated surface would no longer be intrinsically flat, but would have negative or positive curvature, measured by the *deficit angle*, or deviation from 2π encountered when following a closed path. In short, curvature exists on the boundaries between flat simplices. This is intuitively equivalent to the idea that traversing the boundary of a flat paper disk means travelling through an angle of 2π radians from one's starting point, however if a wedge is cut out of the disk and the gap where the wedge was removed is closed up the paper disk is now no longer flat, but must become conical. In other words, the paper disk has acquired some intrinsic curvature, and its circumference has changed from $2\pi r$ to $(2\pi - \theta)r$ where θ is the "deficit angle", as per Fig. 6.2.

In what follows there is a correspondence that can be drawn between triangulations and graphs. A graph is a set of discrete elements often referred to as nodes, points, or vertices, however we will use the term 'hubs', along with a set of associations between some pairs of hubs. These associations are often called edges, lines, or links, however we will use the terminology 'tracks'. Our reason for adopting this slightly unusual terminology is to maintain clarity when switching between discussions of abstract graphs, triangulations, and dual triangulations (to be introduced shortly). If n tracks meet at a hub that hub is said to be n-valent.[2]

---

[1] For instance, one which remains one step away from a specified vertex, as in Fig. 6.2.

[2] In the general case it is possible for a track to begin and end on the same hub, however we will ignore this subtlety for now.

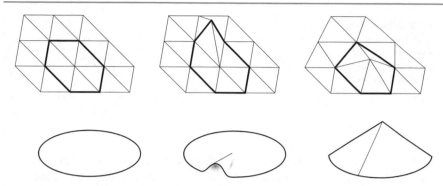

Fig. 6.2 Flat (left) and curved triangulations, compared to a paper disk which is made non-flat by the addition or removal of a wedge-shaped section

A triangulation Δ can be thought of as a type of graph. The vertices of a triangulation correspond with hubs of the associated graph, and edges correspond with tracks. Henceforth any mention of vertices and edges will imply a reference to a triangulation, not an abstract graph.

Given a triangulation Δ, we can construct a dual triangulation, denoted Δ^*, which consists of a graph having hubs at the centres of the simplices of Δ and tracks which each cross through exactly one of the edges of Δ. We shall refer to the hubs of a dual triangulation as 'nodes', and the tracks of a dual triangulation as 'links'. Henceforth any mention of nodes and links will imply a reference to a dual triangulation, not an abstract graph.

In our two-dimensional model, each two-dimensional simplex is dual to a zero-dimensional node, and each one-dimensional edge is dual to a one-dimensional link. Since the simplices in this model are triangular, each node is trivalent (that is, exactly three links meet at each node), as per Fig. 6.3. It is commonplace, and somewhat more general, to refer to each of these elements as k-cells. So a triangular simplex is an example of a 2-cell,[3] edges and links are 1-cells, vertices and nodes are 0-cells. This is the basis of terminology the reader may sometimes encounter, with (dual) triangulations and their higher-dimensional analogues referred to as piecewise linear cell complexes.

As noted above, each 1-cell (edge) in Δ is dual to exactly one 1-cell (link) in Δ^*, and each simplex (2-cell) in Δ is dual to a node (0-cell) in Δ^*. Therefore if a label exists upon an edge, the same label can be associated with the corresponding link, and vice versa.

We can anticipate some of the results to be introduced later by recognising that a graph in which the tracks and/or hubs have some extra information associated with them can model more complex physical systems than a graph which consists of "bare" elements, and these can be relevant to the dynamics of such models. For

---

[3] Though a general 2-cell could be a rectangle or hexagon or any other polygon.

Fig. 6.3 A triangulation
(black) consisting of edges
and vertices, and the
corresponding dual
triangulation (orange)
consisting of links and nodes

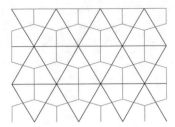

instance Ising models are simply graphs with the value of a spin variable associated
to each hub. In general, the extra information associated to the graph could be a
sense of direction, allowing us to think of hub i as the source of a track, and hub f
as the destination of a track. It could be scalar values corresponding to lengths of the
tracks. In general, it could be elements drawn from any set or group we choose to
utilise. In fact, a bare graph can be thought of as a graph whose tracks are labelled by
copies of the identity element, which are representations of the trivial group. If the
tracks carry some labels, there may be constraints on the labels assigned to adjacent
tracks. We have already encountered an example of such a constraint in the form of
the triangle inequality, as mentioned above.

This brings us to a conception very much like Penrose's original proposal of spin
networks [1]. As conceived, these were graphs with trivalent hubs, each track being
labelled by a positive integer. It was originally intended that these spin networks
could provide a geometrical concept of spacetime structure, and only later was it
recognised that they occurred in theories of quantum gravity. In light of the preceeding
discussion, their similarity to the dual triangulations we have introduced above should
be apparent.

The integer labels n assigned to the tracks of one of Penrose's spin networks must
obey certain constraints. For the three tracks attached to any hub, no single label
should exceed the sum of the other two labels, and the sum of all the labels should be
even. The first of these conditions is simply the triangle inequality. But taken together,
they can be seen to be consistent with the addition of spin angular momentum values,
$j = n\hbar/2$, for some integers $n \geq 0$. Penrose hoped to develop a concept of spacetime
that was not reliant upon the existence of a background coordinate system, and hence
the labels on the tracks of a network could only represent total angular momentum,
not some component thereof in a preferred direction. The rules for coupling the n
values would be used to perform a type of graphical calculation of probabilities,
from which a sense of 'parallelism' and 'orthogonality' could be assigned to pairs
of regions of spin networks. In this way the idea of a geometry for spacetime was
supposed to build up, starting with nothing but combinatorics.

With this in mind we will associate spins $j \geq 0$ with the three links meeting at
a node in Δ^* (and with the corresponding edges in Δ). It is commonly said that we
label the links with representations of SU(2), although this terminology is probably
not the clearest. It would be better to say that we associate representations of SU(2)
with the links and since representations can be labelled by the maximum eigenvalues,
j, which characterise the $(2j + 1)$-dimensional representation of SU(2), we in turn
label the links by the corresponding j values. Recall that SU(2) is a double-cover

of the rotation group in three dimensions, SO(3). We can view the intersection of three links at a node as the coupling of three spins. In general there is an extra detail to be taken into account—the links carry an associated direction, from one node to another. There is then a label associated with each node, called an "intertwining operator" or more commonly just an *intertwiner*, generally denoted ι, which is a mapping from the space of representations ρ_a on the links incoming to that node, to the space of representations ρ_b' on the links outgoing from the node, that is to say that for an $(n + m)$-valent node

$$\iota : \rho_1 \otimes \ldots \otimes \rho_n \to \rho_1' \otimes \ldots \otimes \rho_m' . \tag{6.1}$$

An alternative but equally valid way of regarding an intertwiner is as a mapping from the space of representations that meet at a node, to the complex numbers, $\rho_1 \otimes \ldots \otimes \rho_n \otimes \rho_1' \otimes \ldots \otimes \rho_m' \to \mathbb{C}$. In this case, the intertwiner can be thought of as belonging to the space of products of dual representations, and providing a scalar weighting to each node. This is discussed in more detail in the article by Aquilanti et al. [2].

When first learning quantum mechanics, the addition of angular momentum states $|j_1, m_1\rangle$ and $|j_2, m_2\rangle$ to produce a state $|J, M\rangle$ introduces the concept of Clebsch-Gordon coefficients, which are often written as an inner product $\langle j_1, m_1; j_2, m_2 | J, M\rangle$. However, addition of angular momenta can also equivalently be described with reference to the Wigner 3-j symbols, which are related to the Clebsch-Gordan coefficients through

$$\begin{pmatrix} j_1 & j_2 & j_3 \\ m_1 & m_2 & m_3 \end{pmatrix} \equiv \frac{(-1)^{j_1 - j_2 - m_3}}{\sqrt{2j_3 + 1}} \langle j_1, m_1; j_2, m_2 | j_3, (-m_3)\rangle$$

where the (j_i, m_i) are the orbital and magnetic quantum numbers of the ith system. Although written as 3×2 matrices, these symbols are in fact just scalars. The state $|j_1, m_1; j_2, m_2\rangle$ is the state representing two systems (e.g. particles) each with their separate angular momentum numbers, while $|j_3, m_3\rangle$ represents the *total* angular momentum of the system. Classically, when we have two systems with angular momentum $\vec{L}_1$ and $\vec{L}_2$, the angular momentum of the combined system is $\vec{L}_3 = \vec{L}_1 + \vec{L}_2$. In quantum mechanics, however, the angular momentum of the composite system can be any one of a set of possible allowed choices. Whether or not the angular momentum of the composite system can be specified by quantum numbers j_3, m_3 is determined by whether or not the corresponding Clebsch-Gordan coefficient is non-zero. In contrast, conceptually the 3-j symbols are coefficients of a sum over j and m values such that the linear combination formed from three spin states is zero,

$$\sum_{m_1 = -j_1}^{j_1} \sum_{m_2 = -j_2}^{j_2} \sum_{m_3 = -j_3}^{j_3} |j_1 m_1\rangle |j_2 m_2\rangle |j_3 m_3\rangle \begin{pmatrix} j_1 & j_2 & j_3 \\ m_1 & m_2 & m_3 \end{pmatrix} = |00\rangle . \tag{6.2}$$

The 3-j symbols are non-zero only if three conditions are met; the sum $m_1 + m_2 + m_3 = 0$, the j values are all greater than or equal to zero, and the j values satisfy the triangle inequality. Given the summation in Eq. (6.2) we can see that 3-j symbols treat the three angular momenta more symmetrically than the Clebsch-Gordon coefficients, so that intertwining of representations at a node maps freely between the links chosen to be incoming and outgoing from a given node. Associating the 3-j symbols to a node means that no single j value exceeds the sum of the other two (e.g. $j_1 + j_2 - j_3 \geq 0$) and furthermore that no value is less than the difference of the other two (e.g. $j_3 \geq |j_1 - j_2|$). These conditions hold true under permutation of the indices 1, 2, 3. Ultimately the constraints on the values of the spin labels mean that the areas of the 2-cells can only take discrete values.

Now that we have a two-dimensional model with descriptive terminology in place, let us develop a three-dimensional model. We will want to do this because the space we live in is three-dimensional, and ultimately the model we develop will correspond to a sheet of a spacelike foliation of four-dimensional spacetime, such as we encountered when performing the ADM splitting in Sect. 4.2. The simplices in the two-dimensional case were triangles— polygons with the minimum possible number of edges. In three dimensions the simplices will be tetrahedra—the polyhedra with the minimum number of faces. Each tetrahedron is bounded by four triangular faces, has six edges, and four vertices. Hence if we label the edges (as before) with eigenvalues of representations of SU(2), each simplex is associated to six spins. As before we can construct a dual cell complex, with nodes corresponding to the centres of tetrahedra, and links piercing the triangular faces of the tetrahedra, as depicted in Fig. 6.4b (as the reader may have deduced from the two-dimensional case, in d dimensions a k-cell is dual to a $(d - k)$-cell). The j values associated to edges are not lengths, but we will see in Sect. 6.3 that they are used in a pairwise fashion to define the areas of the faces of tetrahedral simplices occurring in the three-dimensional case.

As in the two-dimensional case there are constraints to be satisfied in constructing a viable cell complex. The triangle inequality places constraints on the allowed edge lengths, for two-dimensional triangulations. However, restrictions on the areas of faces are not enough to uniquely define a tetrahedron. The relationships between them are determined by an extension of the concept of the 3-j symbol, namely the Wigner 6-j symbols. These occur in the addition of three angular momenta to form a fourth (total) angular momentum, and can be written as the sum over products of four 3-j symbols,

$$
\begin{Bmatrix} j_1 & j_2 & j_3 \\ j_4 & j_5 & j_6 \end{Bmatrix} = \sum_{m_1, \ldots, m_6} (-1)^{\sum_{k=1}^{6}(j_k - m_k)} \begin{pmatrix} j_1 & j_2 & j_3 \\ -m_1 & -m_2 & -m_3 \end{pmatrix}
$$
$$
\times \begin{pmatrix} j_1 & j_5 & j_6 \\ m_1 & -m_5 & m_6 \end{pmatrix} \times \begin{pmatrix} j_4 & j_5 & j_3 \\ -m_4 & m_5 & m_3 \end{pmatrix}
$$
$$
\times \begin{pmatrix} j_4 & j_2 & j_6 \\ m_4 & m_2 & -m_6 \end{pmatrix}. \tag{6.3}
$$

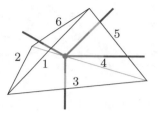
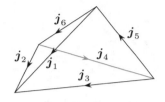

(a) A tetrahedral simplex (black) and its dual (orange)

(b) The tetrahedral Yutsis graph, used to visualise the couplings of angular momenta

Fig. 6.4 Although there are six edges on a tetrahedron, the constraints imposed by the 6-j symbols ensure that only four of these labels are associated with the areas of faces

The form of the product on the right-hand side indicates the couplings of four triples of angular momenta. The 6-j symbol can be visualised as the tetrahedral *Yutsis graph*[4] (Fig. 6.4b). Symmetries of the graph correspond to invariance of the 6-j symbol under permutations of its columns, and the interchange of upper and lower symbols in any two columns. Even though a tetrahedral Yutsis graph has six edges, the 6-j symbol imposes a restriction to a condition between four angular momenta. This can be understood by "peeling open" the tetrahedron into a pair of couplings between four spins, with any three meeting at a vertex of the tetrahedron (corresponding to the decomposition into four 3-j symbols, mentioned above). In the case of Fig. 6.4b we would have, for example, that the value of j_3 is determined by the values of the pairs $\{j_1, j_2\}$ and $\{j_4, j_5\}$, and likewise the value of j_6 is determined by the values of the pairs $\{j_1, j_5\}$ and $\{j_2, j_4\}$. Of course, other permutations of these relationships are equally valid. Hence the constraint requiring that the six edges form a closed tetrahedron ensures that there are only four independent spin labels, which are associated with the areas of faces (or links in the dual cell complex).[5]

The tetrahedral simplices described correspond to tetravalent (i.e. 4-valent) nodes, at which four links l_1, l_2, l_3, l_4 meet. Each node can be decomposed into a pair of trivalent nodes, for instance by considering the links l_1, l_2 and an intermediate link l_i to meet at one node, and the links l_3, l_4 and l_i to meet at the other node. There is, of course, an ambiguity about which external links meet at the same node, resulting in several possible decompositions, as per Fig. 6.5. The intertwining operators at each node provide weightings for each decomposition via the relevant 6-j symbols, which are determined by the 3-j symbols (Fig. 6.6).

Having spent some time developing the conceptual foundations of spin networks, we will now proceed with a more mathematically explicit discussion. Naturally

---

[4] A graph is a Yutsis graph if it can be partitioned into two subgraphs which are each connected (there is a path between each node and any other node) and acyclic (there is only one path between a node and any other node).

[5] For a more detailed discussion of this the reader is recommended to consult Sect. 3.4 of [2], as well as the wikipedia articles '6-j symbol' and 'Racah W-coefficient'.

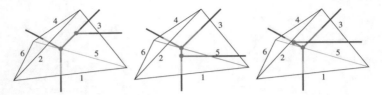

Fig. 6.5 A 4-valent node can be decomposed into a pair of 3-valent nodes in multiple ways. The requirements on the spins meeting at each node, imposed by the 3-j symbols, affect how each decomposition weights the structure of the 4-valent node

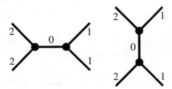

Fig. 6.6 Using Penrose's spin network rules to provide a simplified example of how intertwiners determine the decomposition of a 4-valent node into two 3-valent nodes. The illustration on the left is admissible, but the illustration on the right is not, as one label (with value 2) is greater than the sum of the other two labels at the same node

this will involve reiterating several points already made above, but hopefully the associated ideas will be clearer as a result. Henceforth we will consider a loop state to be a graph or network Γ with tracks labelled by elements of some gauge group (generally SU(2) or SL(2, $\mathbb{C}$) in LQG)

$$\Psi_\Gamma = \psi(g_1, g_2, \ldots, g_m) \tag{6.4}$$

where g_i is the holonomy of A along the ith track, defined in Eq. (6.10), below.[6] Pictorially, we can imagine something like Fig. 6.7a. It is also possible to label the tracks of these graphs by angular momenta, as per Fig. 6.7b, by making use of the Peter-Weyl theorem (Appendix H). At this point a graph Γ consisting of tracks τ labelled by spins j_τ, and hubs χ labelled by intertwiners ι_χ is adopted as the definition of a spin network.[7]

---

[6] The expression "holonomy of a track" may surprise the reader. We have previously referred to a loop holonomy, which measures curvature within a closed path, and described the effect of the connection along a general curve as Eq. (3.26), the Schwinger line integral. As noted in Rovelli and Vidotto [3], the use of the term *holonomy* in the quantum gravity community is a bit different from the more generally-accepted use of the term. For the sake of consistency with other works in the LQG literature we take this opportunity to point out that in LQG the name holonomy is frequently used to refer to the path-ordered exponential of the connection along an arbitrary curve, and will use it in this sense from now on. Only in the special case where the curve is closed upon itself will we revert to using the term "loop holonomy".

[7] The use of χ to denote hubs is somewhat unorthodox. Most authors use the term "vertex" much more freely than is done here, and adopt the symbol v accordingly, but as alluded to above this risks confusing triangulations, dual triangulations, and abstract graphs. As a possibly helpful mnemonic device, recognise that the letter χ looks like the convergence of four tracks at a single hub.

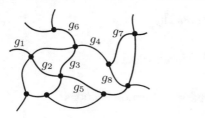
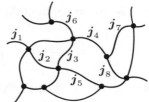

(a) Labelling of the track of a graph   (b) Labelling of the tracks of a
by holonomies   graph by spins

Fig. 6.7 States of quantum geometry are given by arbitrary graphs whose tracks are labelled by group elements representing the holonomy along each track. These graphs can take quite complicated forms. The Peter-Weyl theorem allows us to decompose these states in terms of *spin network* states, where tracks are labelled by group representations (angular momenta)

In general we expect there to be an ensemble $\{(\Psi_\Gamma)_i\}$ of spin networks which corresponds to a semiclassical geometry $\{\mathcal{M}, g_{ab}\}$ in the thermodynamic limit.[8] Shortly we shall identify spin networks with dual triangulations, so that hubs of the graph constituting a spin network are equivalent to nodes, and the tracks of the graph are equivalent to links.

6.2   Space of Generalised Connections

We now wish to identify operators corresponding to physical observables of the spacetime. These operators should be based upon the physical structure of the graphs under consideration. It is worth noting at this point that in the Hamiltonian approach to quantum gravity that we have pursued there is an ambiguity as to whether we choose the connection or the triads as the configuration variables. In fact either choice is permissible, but the physical interpretation of connections as configuration variables and triads as conjugate momenta is more straightforward, and as we shall see it allows us to write operators that generate discrete areas and volumes.

In order to obtain suitable regularised operators in conventional quantum field theory, one must smear the field corresponding to configuration and momentum variables over three-dimensional regions. For instance the operator $\hat{\Phi}_f$, for the configuration variable ϕ in a scalar field theory, would be constructed by smearing the field operator with some function $f(x^\mu)$ over some compact subset $U \in \mathcal{M}$ of the background manifold $\mathcal{M}$,

$$\hat{\Phi}_f = \int_U d^3x \, f(x)\hat{\phi}(x) \tag{6.5}$$

---

[8] When the number of degrees of freedom $N \to \infty$, the volume $V \to \infty$ and the number density $N/V \to n$ where n is bounded from above.

and similarly for the momentum operators $\hat{\Pi}_g$,

$$\hat{\Pi}_g = \int_U d^3x \, g(x)\hat{\pi}(x) \, . \tag{6.6}$$

Given that the local operators satisfy the commutation relations

$$\left[\hat{\phi}(x), \hat{\pi}(x')\right] = i\hbar\delta^3(x, x') \tag{6.7}$$

one can now compute the commutator of the smeared operators:

$$\begin{aligned}
\left[\hat{\Phi}_f, \hat{\Pi}_g\right] &= \int_U d^3x \, d^3x' \left[f(x)\hat{\phi}(x), g(x')\hat{\pi}(x')\right] \\
&= i\hbar \int_U d^3x \, f(x)g(x')\delta^3(x, x') \\
&= i\hbar \int_U d^3x \, f(x)g(x) \, .
\end{aligned} \tag{6.8}$$

The problem with using this prescription for constructing a quantum theory of gravity is that it depends on the structure of the background spacetime, which enters through the integration measure. In a curved spacetime with metric $g_{\mu\nu}$, the integral (6.5) will include a factor of $\sqrt{-\det(g)}$, hence

$$\hat{\Phi}_f = \int_U d^3x \, \sqrt{-\det(g)} f(x)\hat{\phi}(x) \, . \tag{6.9}$$

Our goal is a background-independent treatment of geometrical observables. For this to be possible the smearing procedure should also be background-independent. In a theory of connections and triads such a procedure is already well known—the construction of holonomy variables by integrating ("smearing") the connection along a one-dimensional curve. We write holonomies (compare Eq. (3.30)) in the form

$$g_\lambda[A] = \mathcal{P}\exp\left\{\int_\lambda in_a(x) \, A^a{}_I t^I dx\right\} \tag{6.10}$$

where λ is the curve along which the holonomy is evaluated, x is an affine parameter along that curve, the t^I are generators of the appropriate symmetry group as noted in Sect. 3.1, and n_a is the tangent to the curve at x. What makes holonomies "good" variables for constructing a background-independent theory is the fact that the algebra $\mathrm{Cyl} = \bigcup_\Gamma \mathrm{Cyl}_\Gamma$, (where Cyl_Γ is the algebra of cylindrical functions on the graph Γ, whose elements are of the form (6.4)) constructed on *all possible graphs* on a manifold $\mathcal{M}$ is dense in the space of all suitably regular connections $\mathcal{A}$ on $\mathcal{M}$. In other words, given any connection A on $\mathcal{M}$, by considering all possible graphs on

$\mathcal{M}$, with each track labelled by the holonomy of the connection A along that track, the full gauge invariant information about A can be reconstructed.[9]

So finally we have that the kinematical Hilbert space for a single track τ is $\mathcal{H}_\tau = L^2(\mathcal{G}, d\mu)$—the space of square integrable functions on the *group manifold* of the group $\mathcal{G}$, with $d\mu$ the invariant measure (Haar measure)[10] on the group manifold. For a graph Γ, the kinematical Hilbert space is the tensor product space

$$\mathcal{H}_\Gamma = \bigotimes_\tau \mathcal{H}_\tau \tag{6.11}$$

over all tracks $\tau \in \Gamma$ in the graph.

Given two states of **different** graphs Γ and Γ', their inner product is zero

$$\langle \Theta_{\Gamma'} | \Psi_\Gamma \rangle = \delta_{\Gamma,\Gamma'}. \tag{6.12}$$

Given two different states on the **same** graph Γ, their inner product can be defined using the Haar measure $d\mu$, as

$$\langle \Theta_\Gamma | \Psi_\Gamma \rangle = \int_{\mathcal{G}^m} d\mu_1 \ldots d\mu_m \Theta(g_1, \ldots, g_m) \bar{\Psi}(g_1, \ldots, g_m) \tag{6.13}$$

where m is the number of tracks in the graph.

However, there is an ambiguity in the above procedure because a given state $|\Psi\rangle$ may be cylindrical with respect to more than one graph Γ. This difficulty can be overcome by extending the configuration space $\mathcal{A}$ of regular (smooth, continuous) connections on $\mathcal{M}$ to the space $\bar{\mathcal{A}}$ of *generalized connections* whose elements can be arbitrarily discontinuous and need only be continuous along one-dimensional curves.

For further details and discussion the reader is referred to [6–8].

6.3 Area Operator

The area operator in quantum geometry is defined in analogy with the classical definition of the area of a two-dimensional surface S embedded in some higher dimensional manifold $\mathcal{M}$. To impart an intuitive idea of what follows, we may consider a classical tetrahedron, with each of its triangular faces qualifying as a surface S as just described. If we treat one vertex of this tetrahedron as the origin,

---

[9] For the interested reader, in particular, we recommend reading [4, Sect. 3] and [5, Sect. I.2] for details on the historical developments which led to use of spin networks as the basic objects in LQG.

[10] A Haar measure assigns an invariant concept of volume to subsets of locally compact topological groups (these are sets for which each neighbourhood of a point can be considered to be a closed set, the neighbourhoods around any two distinct points have empty intersection i.e. it is Hausdorff, and the set is endowed with a binary operation making it a group.) The exact form of this measure will depend on the group in question. It serves the useful purpose of establishing a concept of volume that makes it possible to define integrals for the functions of locally compact topological groups.

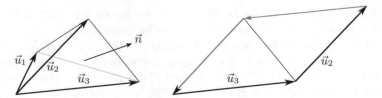

Fig. 6.8 Classically a tetrahedron can be defined by three vectors, with any two defining one of the three faces closest to the origin, and all three defining the fourth face. The magnitude of the wedge product of any pair of vectors is twice the area of the corresponding face, and the length of the normal vector $\vec{n}$ to that face is proportional to this area, since the normal vector and wedge product are dual to each other

then the tetrahedron can be defined by a set of three vectors, denoted $\vec{u}_1$, $\vec{u}_2$, $\vec{u}_3$, all originating at this vertex (as per Fig. 6.8). The other three vertices are located at the tips of the three vectors. Each face can then be thought of as a bivector, $\frac{1}{2}\left(\vec{u}_i \wedge \vec{u}_j\right)$. The normal vector to the face is dual to this bivector, and thus its length measures the area of the face. We immediately recognise that classical angular momentum may be regarded as a bivector $\vec{r} \wedge \vec{p}$, owing to the connection between wedge products and cross products noted in Sect. B.3. The same holds true of angular momentum in quantum mechanics. This hints strongly that the area of a face can be regarded as a measure of angular momentum, and in the quantum case, with links (carrying angular momentum labels) being dual to the triangular faces of tetrahedral simplices, the area of a face should be related to the spin eigenvalues of the associated links. This implies that the areas of the faces would take discrete values. Let us explore these ideas more precisely.

In the simplest case the two-dimensional surface S is a piece of $\mathbb{R}^2$ embedded in $\mathbb{R}^3$. In general both S and the higher-dimensional manifold may have some curvature. To make use of notation developed above, and without loss of generality, we will presume S is embedded in a three-dimensional manifold Σ obtained by foliating four-dimensional spacetime (see Sect. 4.2). To each point $s \in S$ we can associate a triad or "frame field" i.e. a set of vectors $\{\vec{e}_1, \vec{e}_2, \vec{e}_3\}$ which form a basis for the tangent space T_s at that point. In abstract index notation this basis can also be written more succinctly as $\{e_a{}^i\}_s$ where $a, b, c \in \{1, 2, 3\}$ index the vectors and $i, j, k \dots$ label the components of each individual vector in the active or "chosen" coordinate system. The indices $i, j, k \dots$ are necessary because if S is curved (i.e. the gauge connection A_a is non-zero) the basis at two distinct points in S need not be the same, and hence a given vector $\vec{e}_a$ will have different components at different points.

The area of a two-dimensional surface S embedded in Σ is given by

$$A_S = \int d^2x \sqrt{{}^2 h} \tag{6.14}$$

where ${}^2 h_{ab}$ is the metric on S, induced by the three-dimensional metric h_{ab} on Σ, and ${}^2 h$ is its determinant, consistent with Eq. (2.21). Given an orthonormal triad field $\{e_a{}^i\}$ on Σ, we can always apply a local gauge rotation to obtain a new triad basis $\{e'_a{}^i\}$, such that two of its legs—a "dyad" $\{e'_x{}^i, e'_y{}^j\}$—are tangent to the surface S

and $e_z'^k$ is normal to S. Then the components of the two-dimensional metric $^2h_{AB}$ ($A, B \in \{x, y\}$ are purely spatial indices) can be written in terms of the dyad basis $\{e_A{}^I\}^{11}$ as

$$^2h_{AB} = e_A{}^I e_B{}^J \delta_{IJ} . \tag{6.15}$$

The above expression with all indices shown explicitly becomes

$$^2h_{AB} := \begin{pmatrix} h_{xx} & h_{xy} \\ h_{yx} & h_{yy} \end{pmatrix} = \begin{pmatrix} e_x{}^I e_x{}^J & e_x{}^I e_y{}^J \\ e_y{}^I e_x{}^J & e_y{}^I e_y{}^J \end{pmatrix} \delta_{IJ} . \tag{6.16}$$

Now, the determinant of a 2×2 matrix $^2h_{AB}$ takes the well-known form[12]

$$\det(^2h_{AB}) = \sum_{i_1, i_2} h_{1 i_1} h_{2 i_2} \epsilon^{i_1 i_2} = h_{11} h_{22} - h_{12} h_{21} . \tag{6.17}$$

For an orthornormal triad $\epsilon^{ij}{}_k e_z{}^k = e_x{}^i e_y{}^j$. Therefore in terms of the dyad basis $\{e_A{}^I\}$, adapted to the surface S, the expression for the determinant becomes

$$\begin{aligned} \det(^2h_{AB}) &= \left(e_x{}^i e_x{}^j e_y{}^k e_y{}^l - e_x{}^i e_y{}^j e_y{}^k e_x{}^l \right) \delta_{ij} \delta_{kl} \\ &= \left(\epsilon^{ik}{}_m \epsilon^{jl}{}_n - \epsilon^{ij}{}_m \epsilon^{kl}{}_n \right) e_z{}^m e_z{}^n \delta_{ij} \delta_{kl} \\ &= \epsilon^{ik}{}_m \, \epsilon_{ikn} \, e_z{}^m e_z{}^n \\ &= \delta_{mn} \, e_z{}^m e_z{}^n \end{aligned} \tag{6.18}$$

where we have used the fact that $\epsilon^{ij}{}_m \, \delta_{ij} = 0$ and also chosen to write the contraction of two completely anti-symmetric tensors in terms of products of Kronecker deltas.

Thus the classical expression[13] for the area becomes

$$A_S = \int_S d^2x \, \sqrt{\vec{e}_z \cdot \vec{e}_z} \tag{6.21}$$

---

[11] $I, J \in \{0, 1\}$ label generators of the group of rotations SO(2) in two dimensions. They are what is left of the "internal" $\mathfrak{su}(2)$ degrees of freedom of the triad when it is projected down to S.

[12] This is a special case of the determinant for an $n \times n$ matrix A_{ij} which can be written as $\det(A) = \sum_{i_1...i_n \in \mathcal{P}} A_{1 i_1} A_{2 i_2} \ldots A_{n i_n} \epsilon^{i_1 i_2 ... i_n}$ where the sum is over all elements of the permutation group of the set of indices $\{i_m\}$ and $\epsilon^{i_1 i_2 ... i_n}$ is the completely anti-symmetric tensor.

[13] This is only valid for the case when Σ is a three-dimensional manifold. In a general d-dimensional manifold, the area is a tensor

$$A_{\mu\nu}{}^{jk} = e_{[\mu}{}^j e_{\nu]}{}^k . \tag{6.19}$$

In order to extract a single number—the "area"—from this tensor we project onto a two-dimensional surface spanned by the vectors $\{u^j, v^k\}$

$$A[S] = e_{[\mu}{}^j e_{\nu]}{}^k u_j v_k . \tag{6.20}$$

where we suppress the indices μ, ν on the left-hand-side, which distinguish the orientation of the area we are evaluating, just as in the three-dimensional case z is orthogonal to the area we are evaluating, as in Eq. (6.21).

where $\vec{e}_z \cdot \vec{e}_z \equiv e_z{}^m e_z{}^n \delta_{mn}$. With the classical version in hand it is straightforward to write down the quantum expression for the area operator. In the connection representation, the classical triad plays the role of the momenta. Since the quantum operator for the triad is given by $e_a{}^j \rightarrow -i\hbar \frac{\delta}{\delta A^a{}_j}$ we find that

$$\hat{A}_S = \int_S d^2x \sqrt{\delta_{jk} \frac{\delta}{\delta A^z{}_j} \frac{\delta}{\delta A^z{}_k}} . \tag{6.22}$$

In order to determine the action of this operator on a spin network state, let us recall the form of the state Ψ_Γ from Eq. (6.4),

$$\Psi_\Gamma = \psi(g_1, g_2, \ldots, g_m),$$

where g_l is the holonomy along the lth track of the graph. Let the tracks of the graph Γ puncture the surface S at exactly r locations, $\{P_1, P_2, \ldots P_r\}$, as in Fig. 6.9. For the time being let us ignore the cases when a track is tangent to S. We will also ignore the possibility that if the tracks intersect, creating hubs, such a hub happens to lie on S. Then, evidently, the action of Eq. (6.22) on the state Ψ_Γ will give us a non-zero result only in the vicinity of the punctures.[14] Thus

$$\hat{A}_S \Psi_\Gamma \equiv \sum_{p=P_1}^{P_r} \sqrt{\delta_{ij} \frac{\delta}{\delta A^z{}_i(p)} \frac{\delta}{\delta A^z{}_j(p)}} \Psi_\Gamma . \tag{6.23}$$

We have written the connections with an explicit dependence on position p to emphasise that at the lth puncture, the operator will act only on the holonomy g_l. Then recognising that the functional derivative of the holonomy (Eq. (6.10)) with respect to the connection takes the form

$$\frac{\delta}{\delta A^a{}_I} g_\lambda[A] = n_a(x) t^I g_\lambda[A] \tag{6.24}$$

it follows easily that

$$\frac{\delta}{\delta A^a{}_I} \psi(g_1, \ldots, g_k, \ldots, g_m) = n_a t^I \psi(g_1, \ldots, g_k, \ldots, g_m) \tag{6.25}$$

where n^a is the unit vector tangent to the track at the location of the puncture. Now, recall that the t^I in the above expression is nothing more than the Ith generator of the Lie group in question. For SO(3), these generators are the same as the angular momentum operators: $t^I \equiv J^I$. Thus the effect of taking the derivative with respect

---

[14] Since the connection is *defined* only along those edges and nowhere else!.

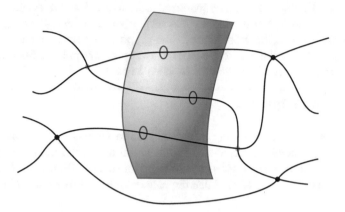

Fig. 6.9 A spin network intersecting a surface at a series of locations ("punctures"). The circles around each puncture are not part of the spin network, but should be interpreted as symbolising that the network endows the surface with quanta of area at these punctures

to the connection is to act on the state by the angular momentum operators. This gives us

$$\frac{\delta}{\delta A^a{}_I} \frac{\delta}{\delta A^b{}_J} \psi = n_a n_b \boldsymbol{J}^I \boldsymbol{J}^J \psi . \tag{6.26}$$

Performing the contractions over the spatial and internal indices, noting that $n^a n_a = 1$, we finally obtain

$$\hat{A}_S \Psi_\Gamma \equiv \sum_k \sqrt{\delta^{ij} \hat{\boldsymbol{J}}_i \hat{\boldsymbol{J}}_j} \Psi_\Gamma = \sum_k \sqrt{\boldsymbol{J}^2} \Psi_\Gamma \tag{6.27}$$

where $\hat{\boldsymbol{J}}_i$ is the ith component of the angular momentum operator acting on the spin assigned to a given track. $\boldsymbol{J}^2$ is the usual Casimir of the rotation group—that is, it is the element $\sum_a X_a X^a$ where the X_a are the basis of the relevant Lie algebra and the X^a are the dual basis defined with respect to some invariant mapping of the basis and dual basis to the scalars. The basic example of a Casimir element encountered at undergraduate level is the squared angular momentum operator $L^2 = L_x^2 + L_y^2 + L_z^2$. Casimir operators commute with all elements of the Lie algebra. The action of $\boldsymbol{J}^2$ upon a given spin state gives us

$$\boldsymbol{J}^2 |j\rangle = j(j+1)|j\rangle . \tag{6.28}$$

This gives us the final expression for the area of S in terms of the angular momentum label j_k assigned to each track of Γ which happens to intersect S,

$$\hat{A}_S \Psi_\Gamma = l_P^2 \sum_k \sqrt{j_k(j_k + 1)} \Psi_\Gamma \tag{6.29}$$

where l_P^2 is inserted in order for both sides to have the correct dimensions.[15] We have therefore found a way of assigning quantised areas to graph states. As the discussion in Sect. 6.1 implied, we would like to use this to operator to assign area to the faces of a simplex. We can therefore identify the graph of a spin network with a dual triangulation, and ascribe area to the faces of a simplex by applying Eq. (6.29) to the spin labels of the links which are dual to these faces.

6.4   Volume Operator

Having found a way of assigning quantised area to the faces dual to the tracks of graph states (spin networks) in Sect. 6.3 it is natural to search for an operator which could assign volume to the regions enclosed by these faces (Fig. 6.10). Similarly to the two-dimensional case, we find that the volume of a region of space $\mathcal{R}$ is given by

$$V = \int_{\mathcal{R}} d^3x \sqrt{h} = \frac{1}{6} \int_{\mathcal{R}} d^3x \sqrt{\epsilon_{abc}\epsilon^{ijk} e_i^a e_j^b e_k^c} \,. \tag{6.30}$$

Replacing the tetrads by their operator equivalents gives us the following expression for the volume *operator*:

$$\widehat{V} = \frac{1}{6} \int_{\mathcal{R}} d^3x \sqrt{\epsilon_{abc}\epsilon^{ijk} \frac{\delta}{\delta A_a{}^i} \frac{\delta}{\delta A_b{}^j} \frac{\delta}{\delta A_c{}^k}} \,. \tag{6.31}$$

We have already discussed in the previous section that the effect of acting on a spin network state with the operator corresponding to the triad has the effect of multiplying the state by the angular momentum operator,

$$n^a \frac{\delta}{\delta A_a{}^i} \Psi_\Gamma = \widehat{J}^i \Psi_\Gamma \,. \tag{6.32}$$

Consequently the action of the volume *operator* on a given state can be expressed as

$$\widehat{V} \Psi_\Gamma = \frac{1}{6} \int_{\mathcal{R}} d^3x \sqrt{\epsilon_{abc}\epsilon^{ijk} n^a n^b n^c \widehat{J}_i \widehat{J}_j \widehat{J}_k} \, \Psi_\Gamma \,. \tag{6.33}$$

Now, since the operator's action is non-zero only on the hubs χ of the graph Γ, the integral in the above expression reduces to a sum over a finite number of hubs $\chi \in \Gamma$ which lie in $\Gamma \cap \mathcal{R}$,

$$\widehat{V} \Psi_\Gamma = \frac{1}{6} \sum_{\chi \in \Gamma \cap \mathcal{R}} \sqrt{\epsilon_{abc}\epsilon^{ijk} n^a n^b n^c \widehat{J}_i \widehat{J}_j \widehat{J}_k} \, \Psi_\Gamma \,. \tag{6.34}$$

---

[15] l_P^2 is also known as the "Planck area", a unit of area given as the square of the Planck length $l_P = \sqrt{G\hbar/c^3}$. We have already encountered this quantity in Chap. 1 when discussing the BH entropy. It should be noted that if the Barbero-Immirzi parameter γ is not real, Eq. (6.29) will also include a factor of γ.

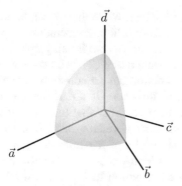
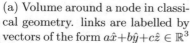

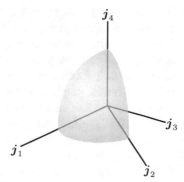

(a) Volume around a node in classical geometry. links are labelled by vectors of the form $a\hat{x}+b\hat{y}+c\hat{z} \in \mathbb{R}^3$

(b) Volume operator in quantum geometry. Links are labelled by elements of the form $\alpha\sigma_x + \beta\sigma_y + \gamma\sigma_z \in \mathfrak{sl}(2, \mathbb{C})$

Fig. 6.10 In order to calculate the volume around a node we must sum over the volume contained in the solid angles between each unique triple of links. Classically this volume can be determined by the usual prescription $\vec{a} \cdot (\vec{b} \times \vec{c})$, where $\vec{a}, \vec{b}, \vec{c}$ are the vectors along each link in the triple. In quantum geometry these vectors are replaced by irreps of SU(2) but the basic idea remains the same

In the literature one finds several forms of the volume operator.[16] Two of these are the Rovelli-Smolin (RS) and Ashtekar-Lewandowski (AL) versions. The RS version [11] is

$$\widehat{V}_{\mathcal{R}}^{\mathrm{RS}}\Psi_\Gamma = \gamma^{3/2}l_\mathrm{P}^3 \sum_{\chi\in\Gamma\cap\mathcal{R}}\sum_{i,j,k} \left|\frac{iC_{\mathrm{reg}}}{8}\epsilon_{abc}\epsilon^{ijk}n^a n^b n^c \hat{J}_i\hat{J}_j\hat{J}_k\right|^{1/2} \Psi_\Gamma \qquad (6.35)$$

where ϵ_{abc} is the alternating tensor, and C_{reg} is a regularization constant.

The AL version [8] is

$$\widehat{V}_{\mathcal{R}}^{\mathrm{AL}}\Psi_\Gamma = \gamma^{3/2}l_\mathrm{P}^3 \sum_{\chi\in\Gamma\cap\mathcal{R}} \left|\frac{iC_{\mathrm{reg}}}{8}\epsilon_\chi(n^a, n^b, n^c)\epsilon_{abc}\epsilon^{ijk}n^a n^b n^c \hat{J}_i\hat{J}_j\hat{J}_k\right|^{1/2} \Psi_\Gamma,$$

$$(6.36)$$

where $\epsilon_\chi(n^a, n^b, n^c) \in \{-1, 0, 1\}$ is the orientation of the three tangent vectors at χ to the three tracks meeting at χ. The term $\epsilon_\chi(n^a, n^b, n^c)$ takes the value zero if the tangents to the tracks are linearly dependent, and if they are independent takes a value reflecting their overall orientation, into or out of the hub. The key difference between the two versions lies in this term. The RS operator does not take account

---

[16] In [9] a construction based on the geometry of classical polyhedra is used to obtain an expression for the volume operator which acts on nodes with valence greater than four. For nodes with fewer than four links, the associated volume always vanishes independent of the choice of the volume operator. For other interesting work, see [10] where the pentahedral volume operator is analyzed and classical chaos is found in the resulting dynamics.

of the orientation of the tracks which meet at the hub, while the AL version does, and it allows us to speak of a phase transition from a high-temperature $(T > T_c)$ state of geometry, where the volume operator averages to zero for all graphs (which are "large" in some suitable sense) and a low-temperature $(T < T_c)$ state where a geometric condensate forms and the volume operator gains a non-zero expectation value for states on all graphs. Temperature, in this case, is a parameter with an externally-imposed value, which determines the probability that the orientation of any track will flip. The key point here is that the AL version takes into account the "sign" of the volume contribution from any triplet of tracks meeting at a hub. Given any such triplet of tracks e_I, e_J, e_K, by flipping the orientation of any one track we flip the sign of the corresponding contribution to $\widehat{V}_S^{AL}$. If we take the orientation of a track as our random variable for the purposes of constructing a thermal ensemble, then it is clear that in the high-temperature limit these orientations will flip randomly and the sum over the triplets of tracks in $\widehat{V}_S^{AL}$ will give zero for most (if not all) graph states. As we lower the temperature the system begins to anneal and for some temperature $T = T_c$ the system should reach a critical point where the volume operator spontaneously develops a non-zero expectation on most (if not all) graph states.

Note that:

1. Since the result of the volume operator acting on a hub depends on the signs $\epsilon(e_I, e_J, e_K)$ of each triplet of tracks, a simple dynamical system would then consist of a fixed graph with fixed spin assignments (j_e) to tracks but with orientations that can flip, i.e. $j_e \leftrightarrow -j_e$ (much like a spin).
2. The Hamiltonian must be a hermitian operator. This fixes the various terms one can include in it. We must also include all terms consistent with all the allowed symmetries in our model.
3. The simplest trivalent spin network has one node with three links (as depicted by the orange hexagonal lattice in Fig. 6.3). One can generalize the action of the volume operator on graphs which have hubs with valence n (number of connecting tracks) greater than 3. (The volume operator gives zero on vertices with $n \leq 2$ so these are excluded). To do so we use the fundamental identity which allows us to decompose the state describing a hub with $n \geq 4$ into a sum over states with $n = 3$. Examples of the decomposition of a four valent hub into two three-valent hubs are given in Fig. 6.5.
4. This model can help us understand how a macroscopic geometry can emerge from the "spin" or many-body system described by a Hamiltonian, which contains terms with the volume and area operators, on a spin network.

We are supposing here that the connections between hubs are fixed in place, to create a static lattice, and only their orientations are allowed to vary. On a tetrahedral lattice with fixed $j = 1/2$, the associated volume at each hub has only two eigenvalues, plus or minus V_0, with the sign being determined by the orientation of the spins on the adjoining tracks.

6.5 Spin Networks

This discussion leaves us with a simple mental picture of spin networks, which we sketched in a somewhat intuitive fashion in Sect. 6.1, and which we reiterate here after having developed it more rigorously. Briefly, spin networks are graphs with representations ("spins") of some gauge group (generally SU(2) or SL(2, $\mathbb{C}$) in LQG) living on each track. We can equate the spin network graphs with dual triangulations of a manifold, with each n-valent node being dual to a simplex with n faces, and each link being dual to a face. By the arguments used above to define area and volume operators, the areas of the faces of a simplex, and the area of the simplex itself are determined by the spin labels on the relevant links. Since each link corresponds with the parallel transport of spin from one node to another, it is necessary to ensure that angular momentum is conserved at nodes, and so an intertwiner is associated with each node. For the case of a four-valent node we have four spins, (j_1, j_2, j_3, j_4). More generally a polyhedron with n faces represents an intertwiner between the edges piercing each one of the faces.

To put it another way, there is a simple visual picture of the intertwiner. In the four-valent case, picture a tetrahedron enclosing a given node, such that each link pierces precisely one face of the tetrahedron. Now, the natural prescription for what happens when a surface is punctured by a spin is to associate the Casimir of that spin J^2 with the puncture. The Casimir for spin j has eigenvalues $j(j+1)$. These eigenvalues are identified with the area associated with a puncture.

In order for the links and nodes to correspond to a consistent geometry it is important that certain constraints be satisfied. For instance, for a triangle we require that the edge lengths satisfy the triangle inequality $a + b < c$ and the angles should add up to $\angle a + \angle b + \angle c = \kappa\pi$, with $\kappa = 1$ if the triangle is embedded in a flat space and $\kappa \neq 1$ denoting the deviation of the space from zero curvature (positively or negatively curved).

In a similar manner, for a classical tetrahedron, now it is the sums of the areas of the faces which should satisfy "closure" constraints. For a quantum tetrahedron these constraints translate into relations between the operators j_i which endow the faces with area.

For a triangle, giving its three edge lengths (a, b, c) completely fixes the angles and there is no more freedom. However, specifying the areas of all four faces of a tetrahedron *does not* fix all the freedom. The tetrahedron can still be bent and distorted in ways that preserve the closure constraints. These are the physical degrees of freedom that an intertwiner possesses—the various shapes that are consistent with a tetrahedron with a given set of face areas. We can understand this as meaning that when tetrahedra are "glued together" the areas of the joined faces must match, but the shapes of the joined faces do not need to match (and are, in fact, meaningless in a background-independent context).

6.6   Looking Ahead to Spin Foams

In LQG the kinematical entities describing a given state of quantum geometry are spin networks. The dynamical entities—i.e. those that encode the evolution and history of spin networks—are known as *spin foams*. If a spin network describes a d-dimensional spacelike geometry, then a spin foam describes a possible history which maps this spin network into another one, via some series of intermediate spin networks embedded within $d + 1$ dimensions, all of which are within the kinematical Hilbert space. The name spin foam derives from the superficial resemblance borne to a mass of soap bubbles by the structure produced when a spin network is "extruded" through one higher dimension. In order to determine the transition amplitudes between two different states of quantum geometry whose initial and final states are given by spin networks S_i and S_f, one must sum over *all* possible spin foams which interpolate between the two spin network states. This sum over spin foams should sound familiar, since it is essentially the sum over geometries mentioned near the end of Sect. 4.1 in our discussion of the Einstein-Hilbert action, and subsequent discussion of the ADM splitting.

We will turn our attention to spin foams more fully in the next chapter, but note here that when the sum over all allowed histories is performed one finds that the resulting amplitude depends only on the *boundary* configuration of spins. This is holography. The holographic principle boils down to saying that the state of a system is determined by the state of its boundary. Therefore, although the point is not made as often as it possibly should be, LQG embodies the holographic principle in a very fundamental way. This point is discussed in more detail in Chap. 9.

The remainder of this book will deal with the dynamics of spacetime in loop quantum gravity, and the applications of LQG to physical scenarios which may lend themselves to experimental testing of the theory. It is not our intention to discuss any of the subsequent topics in exhaustive detail. We do hope, though, that a good overview of the recent developments in LQG, and its applications to a range of topics (like cosmology, for instance) will help anyone new to the field understand where to turn next as they investigate loop quantum gravity and related ideas for themselves.

References

1. R. Penrose, Angular momentum: an approach to combinatorial space-time. *Quantum Theory and Beyond* Ed. by T. Bastin (1971), pp. 151–180. https://math.ucr.edu/home/baez/penrose/
2. V. Aquilanti et al., Semiclassical mechanics of the Wigner 6j-symbol. J. Phys. A: Math. Theor. **45**, 065209 (2012). https://doi.org/10.1088/1751-8113/45/6/065209, arXiv:1009.2811v2
3. C. Rovelli, F. Vidotto, *Covariant Loop Quantum Gravity: An Elementary Introduction to Quantum Gravity and Spinfoam Theory*. (Cambridge University Press, 2014). ISBN: 9781107706910. https://doi.org/10.1017/CBO9781107706910
4. A. Ashtekar, J. Lewandowski, Background independent quantum gravity: a status report. Class. Quant. Grav. **21**(15), R53–R152 (2004). https://doi.org/10.1088/0264-9381/21/15/R01. arXiv:gr-qc/0404018. url: http://dx.doi.org/10.1088/0264-9381/21/15/R01
5. T. Thiemann, Introduction to Modern Canonical Quantum General Relativity (2001). https://doi.org/10.48550/gr-qc/0110034. arXiv:gr-qc/0110034

6. A. Ashtekar, J. Lewandowski, Differential Geometry on the Space of Connections via Graphs and Projective Limits. J. Geom. Phys. **17**, 191–230 (1995). https://doi.org/10.48550/hep-th/9412073. arXiv:hep-th/9412073. url: http://arXiv.org/abs/hep-th/9412073
7. A. Ashtekar, J. Lewandowski, Quantum theory of gravity I: area operators. Class. Quant. Grav. **14**, A55–A82 (1997). https://doi.org/10.48550/gr-qc/9602046. arXiv: gr-qc/9602046
8. A. Ashtekar, J. Lewandowski, Quantum theory of geometry II: Volume operators. Adv. Theor. Math. Phys. **1**, 388–429 (1998). https://doi.org/10.48550/gr-qc/9711031. arXiv: gr-qc/9711031
9. E. Bianchi, P. Dona', S. Speziale, Polyhedra in loop quantum gravity. Phys. Rev. D **83**, 044035 (2011). https://doi.org/10.1103/PhysRevD.83.044035. arXiv:1009.3402
10. H.M. Haggard, Pentahedral volume, chaos, and quantum gravity. Phys. Rev. D **87**, 044020 (2013). https://doi.org/10.1103/PhysRevD.87.044020. arXiv:1211.7311
11. C. Rovelli, L. Smolin, Discreteness of area and volume in quantum gravity. Nucl. Phys. B **442**, 593–622 (1995). https://doi.org/10.1016/0550-3213(95)00150-Q. arXiv:gr-qc/9411005

Dynamics of Spin Networks

7

The discussion has reached a point where we understand the basic structure of space-time in loop quantum gravity. In short, the similarities between general relativity and quantum field theory have been noted, and we have seen how a foliation that allows us to view spacetime in terms of generalised coordinates and canonical momenta can be constructed. The quantization of this model brings us to a point where the structure of spacetime is viewed as discrete, and composed of intersecting loops. This enables us to view the kinematics of the theory in terms of one-dimensional links that meet at nodes, the valence of any node being equal to the number of links that meet there, and each node being associated with a discrete volume contribution to a spatial manifold. The earlier "chain-mail" conception of interlinking loops (which inspired the name *loop quantum gravity*) has largely given way to this graph-based concept of spin networks.

Of course, the most significant feature of general relativity is that it is a theory of dynamical spacetime. Therefore, while kinematics are important, it is not sufficient to treat spacetime as a static network, but we should also ask how this quantised spacetime can change its geometry. Without this ability, our quantum spacetime is unable to account for the full range of phenomona that occur within general relativity.

To this end, we will now turn to the topic we touched on at the end of the previous chapter, spin foams. We have come a long way since the discussions of quantum field theory and classical general relativity at the start of this book. These topics are covered in a profusion of textbooks, and should be familiar to most physicists, but we are now heading into territory where less familiar concepts lurk, and ideas from group theory, quantum field theory, and graph theory mix together in surprising ways. As we proceed, we will focus on describing concepts clearly and spelling out the meanings of terminology that may be used in unfamiliar ways, and also mention the historical development of these ideas, setting the reader up to explore more thorough mathematical coverage of each topic in their own further reading.

© Springer Nature Switzerland AG 2024
S. Bilson-Thompson, *Loop Quantum Gravity for the Bewildered*,
https://doi.org/10.1007/978-3-031-43452-5_7

7.1   Spin Foams

A spin foam corresponds to a history which connects two spin network states. To build an intuition for how this can be achieved, we will return to the two-dimensional and three-dimensional toy models we developed in Sect. 6.1. Consider a spin network— that is, a dual triangulation Δ^* of a two-dimensional surface S, embedded in three dimensions. Now extend each k-cell in Δ^* away from the surface. Every k-cell will extrude to a $(k+1)$-cell, and the final spin network obtained is a copy of the dual triangulation Δ^*. The cell complex we have obtained is a trivial example of a spin foam. If we think of the spin foam as a mapping from a dual triangulation of an initial surface to a dual triangulation of a final surface then we have in fact just described the "identity spin foam". Although simple, this example serves two primary purposes. Firstly, it explains the motivation behind the name, since it looks a bit like a foam of soap bubbles. And secondly, it shows that the labels on the k-cells of the original triangulation (for $k \geq 0$) are carried by the corresponding $(k+1)$-cells of the spin foam—hence the 2-cells of the spin foam (which we will call sheets) carry spins, and the 1-cells of the spin foam (which we will call edges) carry intertwiners. Naturally we can imagine an equivalent process happening simultaneously with the triangulation Δ, since we can construct Δ from Δ^* and vice-versa, and so we are also constructing a copy of the surface S. The three-dimensional region "swept out" by the triangulation Δ (as opposed to the spin network) as it extends away from S can be identified as $S \times [0, 1]$, where naturally $[0, 1]$ is a subset of $\mathbb{R}$. We can therefore imagine that we have two parallel surfaces, S_0 and S_1. If the process were repeated n times we would have a succession of two-dimensional manifolds S_i, with spin networks embedded in them. This is entirely analogous to the foliation of spacetime which we encountered in Sect. 4.2, and so we should regard the variable i as a time parameter.

Any valid (non-trivial) dynamics for spin networks must transform an initial network in the kinematical Hilbert space into another network also in the kinematical Hilbert space. There are two basic transformations from which the transitions between network states can be built. These are the "2-to-2" move, in which we take a pair of triangular 2-cells connected along a single edge, and redefine the edge to run between the two vertices which they do not initially have in common, and the "1-to-3" move, in which a single triangle is split into three. These evolution moves (and thus the process of extruding a triangulation) can also be viewed in terms of the dual triangulation Δ^*, which is a trivalent graph as noted before. From this perspective the 2-to-2 move consists of an adjacent pair of trivalent nodes in a network exchanging one incoming link each, and the 1-to-3 move consists of a single trivalent node splitting into three, in analogy to the "star-triangle transformation" used in the analysis of electrical circuits (see Fig. 7.1a). The inverse moves are also possible, of course, and the reader should easily recognize that the 2-to-2 move is its own inverse, while the inverse of the 1-to-3 move shrinks a trio of nodes down to form a single node.

It can be seen from Fig. 7.1b (right) that the 1-to-3 move acting on the dual triangulation creates a set of sheets which form a tetrahedron having one vertex in

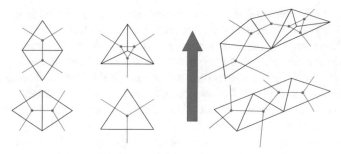

(a) The 2-2 and 1-3 moves in a triangulation (black) and the dual triangulation (orange)

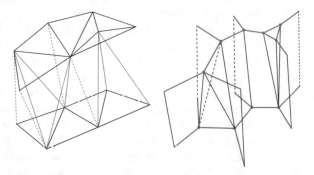

(b) Evolution of a triangulation (left) and dual triangulation (right) through successive spatial hypersurfaces.

Fig. 7.1 The spin networks undergo a succession of evolution moves as time passes, from bottom to top as indicated by the vertical arrow, creating "spin foams". For clarity we restrict ourselves in this case to two-dimensional spatial hypersurfaces. The evolution of d-simplices in the triangulation (triangles in this case) creates a triangulation of the resultant spacetime by $(d+1)$-simplices (tetrahedra, in this case). However, extrusion of one-dimensional links in the dual triangulation creates two-dimensional sheets, which don't necessarily correspond to the faces of $(d+1)$-simplices

the surface S_i and three vertices in the surface S_{i+1}. Likewise the 2-to-2 move acting on the dual triangulation creates a set of sheets which form a tetrahedron having two vertices in each of the surfaces S_i and S_{i+1}. As a result, in what follows, the *nodes* of a spin network embedded in d-dimensional space will sometimes be equated with the *vertices* of a $(d+1)$-simplex. However, it is not necessarily true that the sheets resulting from the evolution of a spin network in S_i to a spin network in S_{i+1} will form simplices of the resulting spacetime. For instance, the trivial identity spin foam described at the start of this chapter does not divide the manifold into d-dimensional simplices. It is worth stating clearly then, to avoid confusion in what follows, that the spin networks are dual triangulations of a d-dimensional manifold, and the spin foams that arise from them can be seen as subdivisions (*rather than* dual triangulations) of a $(d+1)$-dimensional manifold. However, we can see from Fig. 7.1b (left) that a triangulation (as opposed to a dual triangulation) of a two-dimensional surface gives rise to a (2+1)-dimensional structure of tetrahedra. This is the basis of the Regge model of 2+1 gravity, and is fleshed out further in Appendix I.

It should be kept in mind that referring to a spin foam as a two-dimensional piecewise linear cell complex does not mean the spin foam is two-dimensional, or able to be embedded in a two-dimensional manifold. Rather, it means that the fundamental pieces of which the spin foam is built up are two-dimensional. Just as the spin network is a complex of one-dimensional line segments, the spin foam is a complex of two-dimensional sheets.[1]

It is fairly easy to see that a cross-section through any spin foam will single out a spin network intermediate between the initial and final network states. Naturally this network will be of one dimension lower than the foam, hence a four-dimensional spin foam will (when "sliced through" by spacelike three-dimensional hypersurfaces) yield intermediate three-dimensional spin networks.

This is essentially identical to the construction of a spacetime foliation at a fixed value of fiducial time which we have already encountered. Note that the topology of the spin network (i.e. the number of nodes, and the links defined between any pair of nodes) remain unchanged for slices between vertices of the spin foam, but can change when moving between slices that occur "before" and "after" any vertex.

The discussion so far has dealt with spin networks embedded in two-dimensional hypersurfaces, which extrude along the time direction to produce three-dimensional spin foams. Since the universe in which we live is $(3 + 1)$-dimensional we will want to extend the concepts we've developed so far to three-dimensional hypersurfaces which foliate four-dimensional spacetime. Thus we are re-treading the path laid out in Sect. 6.1, and as we argued there, the basic three-dimensional simplices will be tetrahedra. In analogy to the 2-2 and 1-3 moves already encountered, we can consider the evolution moves which carry a spin network in a three-dimensional spatial hypersurface into another hypersurface. One such move is the 1-4 move, which expands a single node in the spin network into a tetrahedral arrangement of four nodes (each node being dual to a tetrahedral 3-simplex). We can therefore see that four-dimensional spin foams will be composed of 4-simplices having five vertices (which we will identify with spin network nodes labelled $1, \ldots, 5$ as in Fig. 7.2, where one node resides in hypersurface S_i and the other four in the surface S_{i+1}), ten edges (that is, the six edges of a tetrahedron, plus another four edges between the fifth node and the other four), and five tetrahedral "faces" (namely the tetrahedra defined by the nodes $\{1, 2, 3, 4\}$, $\{2, 3, 4, 5\}$, $\{3, 4, 5, 1\}$, $\{4, 5, 1, 2\}$, and $\{5, 1, 2, 3\}$). The inverse move, that is the 4-1 move, shrinks a tetrahedral cluster of 3-simplices down to a single node (i.e. a single 3-simplex). We are also led to consider the 3-2 move (or 2-3 move) in which three of the five vertices of a 4-simplex reside in one hypersurface, and the other two reside in the subsequent (or previous) hypersurface. These evolution moves are illustrated in Fig. 7.2a. It is

---

[1] We note that we are restricting the discussion somewhat here for the sake of simplicity. We recommend Sect. 1 of [1] as a good starting-point to investigate more deeply the details around defining spin foams.

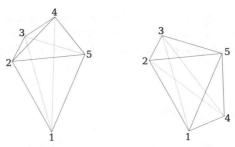

(a) The evolution of spin networks creates four-dimensional simplices between spatial hypersurfaces

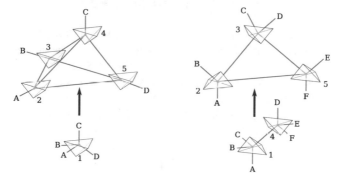

(b) The effect of the 1-4 and 2-3 moves on spatial simplices

Fig. 7.2 Three-dimensional spin networks give rise to four-dimensional spin foams. **a** In the 1-to-4 move (left), node 1 exists in a spatial hypersurface and nodes 2, 3, 4, and 5 exist in the subsequent hypersurface. In the 2-to-3 move (right), nodes 1 and 4 exist in the earlier hypersurface, and nodes 2, 3, and 5 exist in the later hypersurface. **b** The same moves with the structure of simplices made explicit. Corresponding external links in the earlier and later hypersurfaces are identified with matching letters

hopefully clear from this discussion that the tetrahedral arrangement of four nodes in S_{i+1} (arising from a 1-4 move acting on a single node in hypersurface S_i) is not itself a spatial simplex (i.e. 3-simplex), but rather a tetrahedral cluster of spatial simplices, as illustrated in Fig. 7.2b.

As was suggested above by the initial example of an identity spin foam, we can think of spin foams as mappings or operators[2] which act upon the kinematical states to produce new states. Two spin foams that map between dual triangulations[3] Δ_a^* and Δ_b^*, and between Δ_b^* and Δ_c^* can be composed to form a spin foam that maps between Δ_a^* and Δ_c^*, thus $\mathcal{F}_1 : \Delta_a^* \to \Delta_b^*$ and $\mathcal{F}_2 : \Delta_b^* \to \Delta_c^*$ can be composed with

---

[2] A mathematician would use the term *morphism*.

[3] I.e. Between spin networks which represent the states of a spacelike hypersurface at two different times.

the result $\mathcal{F}_2\mathcal{F}_1 \equiv \mathcal{F} : \Delta_a^* \to \Delta_c^*$. Even though we speak here of dual triangulations, implying a background manifold which is broken up into simplices, we are entitled to think of the spin networks these mappings act upon as abstract graphs, independent of any background.

7.1.1 Transition Amplitudes

With a clear mental picture of spin foams in both 2+1 dimensions and 3+1 dimensions hopefully established, one can formally view a spin foam as a succession of states $\{|\Psi(t_i)\rangle\}$ obtained by the repeated action of the scalar (or Hamiltonian) constraint

$$|\Psi(t_{i+1})\rangle \sim \exp\{-i\mathcal{H}_{\text{EHA}}\delta t\}|\Psi(t_i)\rangle \tag{7.1}$$

and so on, where $\mathcal{H}_{\text{EHA}}$ is defined in Eq. (5.3), with $|\Psi(t_0)\rangle$ as the initial state [2,3]. At each node the Hamiltonian constraint acts to modify the geometry of the spin network according to the 2-to-2, 1-to-3, and 3-to-1 moves.

Initial and final spin network states can be viewed as describing configurations of the metric of initial and final spacelike hypersurfaces forming the boundary of a spacetime manifold, $[g_i]$ and $[g_f]$. As a result, the scalar product on the space of metrics

$$\langle[g_i]|\mathcal{P}|[g_f]\rangle = \int_M \mathcal{D}[g]e^{iS_{\text{EH}}} \tag{7.2}$$

is a quantity of particular interest. As this is a measure of how much an initial and final state of geometry overlap, it would be expected to determine the amplitude for transition between initial and final spatial metrics. In this expression we integrate over the space of metrics up to diffeomorphism, which are consistent with the boundary conditions (i.e. $[g_i]$ and $[g_f]$), with $\mathcal{P}$ being the projector on the kernel of the Hamiltonian constraint.

It is the picture of spinfoams as operators, acting upon states which are spin networks, which connects to the sum-over-geometries concept invoked in Chap. 4. The action, and hence the weighting associated with the resultant geometries, is determined by the spin and intertwiner labels carried by the links and nodes. A particular spin foam $\mathcal{F}$ is the entire complex of sheets extruded between the links of intermediate spin networks, the edges between sheets (extruded from nodes of spin networks), and the vertices where these meet. The initial and final spin network states therefore constitute the boundary of $\mathcal{F}$. But the piecewise linear cell complex of a spin foam is adorned with spins and intertwiners inherited from the spin networks it embodies. As a result, to each particular spin foam can be associated a transition amplitude of the form

$$A(\mathcal{F}) = \sum_{j_s,\iota_e} \prod_s A_s(j_s) \prod_e A_e(j_{s_e}, \iota_e) \prod_v A_v(j_{s_v}, \iota_{e_v}), \tag{7.3}$$

where s refers to the "sheets",[4] e refers to the edges between sheets, and v refers to the vertices where these edges meet. Here the notation j_{s_v}, ι_{e_v} refers respectively to the spins labelling the sheets that meet at vertex v and the intertwiners labelling the edges that meet at v. Similarly j_{s_e} refers to the spins labelling the sheets that meet at edge e.

As mentioned, each of these expressions is purely formal, and it is desirable to move beyond a formal view of spin foams, to make explicit calculations feasible. Naturally this is easier said than done, and new approaches continue to be developed. For the remainder of this chapter we will look at some of the historical and current efforts to address this task.

7.2   Early Developments

The first topic we will look at, BF theory, is over twenty years old at the time of writing. In a sense then we are stepping backwards in time, before surveying newer developments which may be of greater interest to the modern reader. However this topic is noteworthy for establishing concepts and terminology which continue to be important in discussions of loop quantum gravity and spinfoams, and furthermore its mathematical formulation is intimately linked to the conceptual image of spin foams laid out in Sect. 7.1.

The discussion of classical GR in Chap. 4 used the Lagrangian and Hamiltonian frameworks to motivate our definitions and concepts. Discussions of BF theory often involve the language of differential forms and exterior derivatives (Appendix B), and so a comparison between the discussions in Chap. 4 (involving indices on terms such as the connection and curvature) and the index-free notation of differential forms may also benefit the reader by providing multiple perspectives. In particular this dovetails with the Palatini formulation which we covered in Sect. 4.3.5. Although not essential for the following discussion, the content of Appendix J provides a guide to some of the language and concepts frequently encountered in the wider literature on these topics.

7.2.1   BF Theory

BF theory has long been of interest as the three-dimensional version contains (2+1) general relativity, however for brevity we will concentrate on the four-dimensional case, in which we obtain (3+1) general relativity as a BF theory with extra constraints. By this point, the mention of constraints should give a hint as to where the discussion is leading.

At the classical level we choose a d-dimensional manifold $\mathcal{M}$ and a gauge group G with an associated Lie algebra $\mathfrak{g}$. The theory involves two basic fields. The first is a 1-form A which we identify as a connection. The second is a $(d-2)$-form which

---

[4] Other authors may refer to them as faces.

we denote E. In four dimensions this obviously makes E a 2-form. Since the exterior derivative of a k-form is a $(k + 1)$-form, the curvature of A is a 2-form which we refer to as F. The choice of symbol here is intentionally reminiscent of the field strength tensor $F_{\mu\nu}$, which we will recall embodied the curvature of the gauge field A_μ in quantum field theory (as per Eqs. (3.15) and (3.16)). In what follows we will make use of d_A, the exterior covariant derivative with respect to the connection A such that $d_A\omega = d\omega + A \wedge \omega$, and recognise the curvature form as $F = dA + A \wedge A$. This bears a clear resemblance to the form of the field strength tensor given in Eq. (3.16). The relevant Lagrangian is obtained by taking the wedge product $E \wedge F$, which defines a d-form, i.e. a form having the same number of dimensions as the manifold $\mathcal{M}$, and taking the trace[5] to obtain

$$\mathcal{L} = \text{Tr}(E \wedge F). \tag{7.4}$$

Notice the similarity to the integrand of the Einstein-Hilbert action in the absence of matter, Eq. (4.64), where we have a curvature term given by $R = g^{\mu\nu}R_{\mu\nu}$, and recall that the curvature term in the BF action is a 2-form. The similarity here becomes even more pronounced when we recall the discussion of tetrads and their relationship to the metric in Sect. 4.3.2. The "extra constraints" mentioned above in fact refer to writing E in terms of the tetrads. Specifically, if we require that $E = e \wedge e$ we can view the E as *simple* bivectors, corresponding to the triangular faces of tetrahedra, as described at the start of Sect. 6.3. As one would expect of a bivector, when the orientation of a face is reversed, the sign of E is reversed. We also require that these tetrahedra have non-zero volume, and that the four faces must meet up properly so that the tetrahedron is closed. The full list of these constraints can be found in [5], and as just mentioned, imposing them ensures that we obtain general relativity. The variation of the action obtained from Eq. 7.4 is zero when

$$\delta \int_{\mathcal{M}} \text{Tr}(E \wedge F) = \int_{\mathcal{M}} \text{Tr}(\delta E \wedge F + (-1)^{n-1}d_A E \wedge \delta A) = 0$$
$$\therefore F = 0, \quad d_A E = 0. \tag{7.5}$$

This result is to be expected in a theory of spacetime without matter—it simply says that the connection is flat. We will notice that $d_A F = 0$ allows for transformations $E \to E + d_A\eta$, adding an appropriately-chosen derivative term to E, which result in physically equivalent solutions [4].

The classical phase space of BF theory is investigated by foliating spacetime, as was done in Sect. 4.2, so that $\mathcal{M} = \Sigma \times \mathbb{R}$. By a suitable choice of gauge the time component of A can be set to zero and we find that the field E becomes the canonically conjugate momentum to A, hence $E = \partial\mathcal{L}/\partial\dot{A}$. The analogy with classical electromagnetism leads to E being referred to as the electric field in some instances (even though the student of quantum gravity may wonder what electric fields are doing in

---

[5] For a more detailed description of *how* to take this trace, see [4].

a theory of dynamical spacetime). It is also responsible for the naming of the Gauss constraint (first mentioned in Sect. 4.2.1), as the condition $d_A E = 0$ is reminiscent of Gauss' law from classical electromagnetism in the absence of charges.[6]

Based upon BF theory Barrett and Crane [5] developed, with notable contributions from Baez, a model of spacetime dynamics which is closely related to spin foams. In this model the spin networks are labelled by representations of SO(4) (or SU(2)×SU(2)), and the tetrahedral 3-simplices extrude into 4-simplices, exactly as we have already discussed. The choice of SO(4) was determined by the desire to shift from a three-dimensional description of spacetime, i.e. (2+1) general relativity, to a four-dimensional description. It was realised [6] that this model could be interpreted in terms of a Feynman graph of a class of theories that came to be known as group field theories, to which we will turn in Sect. 7.3.3. However, the model is not identical to loop quantum gravity [7], and so we will mention it for historical context, but not dwell on it.

7.2.2   Chern–Simons Theory

The image we have built up in the previous sections is of the configurations of simplices (or the appropriate dual cell-complexes) as a kind of Feynman diagram for spacetime structure. Correspondingly, we expect a transition between an initial spin network state and a final spin network state to involve sums over the appropriately weighted intermediate spin foams. We naturally want the sum over spin foams to be triangulation independent, and we can be confident this is the case if we can relate the triangulations to some set of topological invariants.

Frequently when summing over intermediate states in quantum field theory, we encounter divergences which must be tamed somehow. Similarly, when summing over spin foam states we encounter divergences. Roughly speaking, the sheets of the spin foam (being extruded from links in a spin network) are labelled by spins. Naively the sum over possible spin foams can include arbitrarily large spins, leading to transition amplitudes which diverge. What we would like, then, is a way to limit the permissible spin values so that the sum-over-geometries does not diverge. Fortunately in BF theory there is a way to do this, by adding extra terms to the Lagrangian.

In the three-dimensional case the BF action is written in terms of E, F, and a scalar constant Λ as

$$\int_{\mathcal{M}} \text{Tr}\left(E \wedge F + \frac{\Lambda}{6} E \wedge E \wedge E \right). \tag{7.6}$$

---

[6] In earlier works the symbol B was more frequently used for the bivector fields, hence the name BF theory.

This leads to new equations of motion, found by setting the variation of the resultant action to zero, hence

$$F + \frac{\Lambda}{2} E \wedge E = 0, \quad d_A E = 0. \tag{7.7}$$

In this case spacetime is no longer flat—F becomes non-zero.

To see why this is an effective way of eliminating divergences we will need to digress again, to talk about Chern–Simons theory (which is also discussed in Appendix G). This is a background-free gauge theory, originally formulated in three dimensions, with action

$$S_{CS} = \frac{k}{4\pi} \int_M \mathrm{Tr} \left(A \wedge dA + \frac{2}{3} A \wedge A \wedge A \right). \tag{7.8}$$

In fact, Chern–Simons theory is a topological field theory, meaning the observables of the theory tell us about the topological invariants of spacetime itself. The action S_{CS} is (perhaps unsurprisingly, then) a topological invariant.

If we define two connections

$$A_\pm = A \pm \beta E \tag{7.9}$$

then the BF action can be written as the difference of two Chern–Simons actions,

$$S_{CS}(A_+) - S_{CS}(A_-) = \int_M \mathrm{Tr} \left(E \wedge F + \frac{\Lambda}{6} E \wedge E \wedge E \right). \tag{7.10}$$

by appropriate choices of β and k, and so if we can induce a cutoff for spin values in Chern–Simons theory, we should achieve the same for BF theory. The quantity k is referred to as the Chern–Simons level, and will play an important role in taming the divergences that have motivated these steps.

In the Abelian case the wedge product $A \wedge A \wedge A$ in Eq. (7.8), which we can think of a self-interaction term for the connection, is zero and so we are left with the leading $A \wedge dF$ term only, corresponding to a flat connection, which we looked at in Sect. 7.2.1.

The non-Abelian case corresponds to non-zero self-interactions, and results in the $E \wedge E \wedge E$ term in the BF action. The coupling strength is written as Λ, since it is generally identified with a cosmological constant, which we can think of as representing the energy density of empty space. The introduction of this cosmological constant term leads to a modification of the spin networks we have by now come to know and (presumably) love, such that they may be visualised as networks of ribbons. It would be a significant digression to explain the details properly, and so we will provide below a conceptual outline of the reasoning which brings us to this view. For more details we recommend the reader consults [8,9].

7.2.3   The Cosmological Constant

In discussing BF theory above it was mentioned that gauge transformations were possible, and involved adding an appropriate derivative term to E. Working with explicit indices again, to emphasise the similarity to the discussion in Sect. 3.1, it is natural to ask whether the action Eq. (7.8) is invariant under gauge transformations,

$$A_\mu \to A'_\mu = g^{-1}A_\mu g + g^{-1}\partial_\mu g \tag{7.11}$$

(notice the similarity here to Eq. (3.5), though with a slight change in notation). It is here that we must address the concept of winding numbers. Consider a mapping $f(\theta) = e^{i\theta}$. Clearly this is a mapping from the circle to itself, $f : S^1 \to S^1$. We can construct other mappings by taking powers of f, hence $f^w(\theta) = e^{iw\theta}$, but in this case the mapping winds w times around the circle, and w is referred to as the winding number of the mapping. We can also define a function $f_a(\theta) = e^{i(w\theta + a\theta_0)}$. Setting $a = 0$ yields $f_0(\theta) = e^{i(w\theta)}$, while setting $a = 1$ yields another mapping with the same winding number, $f_1(\theta) = e^{i(w\theta + \theta_0)}$. Varying the value of a generates a class of mappings between f_0 and f_1 with the same winding number. Such a class of mappings that can be smoothly deformed into each other is called a *homotopy class*.

What does this have to do with gauge transformations? We can readily identify the mappings $f_a(\theta) = e^{iw\theta + a}$ as U(1) transformations. It is not difficult to generalise to other groups, such as SU(2), as we did in Eq. (3.13). We will refer to the relevant transformations as g, as per Eq. (7.11). In the case of SU(2) this is equivalent to a mapping $g : S^3 \to S^3$, as a Euclidean n-dimensional space with points at infinity identified is equivalent to an n-sphere (the first S^3), while the group manifold of SU(2) is the second S^3.

Applying the gauge transformation above to A_μ and taking the variation of the Chern–Simons action we obtain a sum of two terms

$$\delta S_{CS} = \frac{k}{4\pi} \int_{\mathcal{M}} d^3 x \epsilon^{\mu\nu\rho} \Big(\partial_\mu \text{Tr}(\partial_\nu g)(g^{-1}A_\rho)$$

$$+ \text{Tr} \frac{1}{3}(g^{-1}\partial_\mu g)(g^{-1}\partial_\nu g)(g^{-1}\partial_\rho g) \Big). \tag{7.12}$$

The second term is actually proportional to the winding number w associated to the gauge transformation. As mentioned above, these kinds of mappings can be assigned to distinct homotopy classes. Since gauge transformations associate a group element $g \in G$ with each point, $x \mapsto g(x)$, one possible gauge transformation is the mapping to the identity element. This leads to a distinction between mappings that can be connected to the identity, and those that cannot. A transformation corresponding to anything less than a full rotation has a winding number of zero and can be deformed to no winding. These are generally called 'small' gauge transformations. Transformations corresponding to a full rotation, or several full rotations, have non-zero winding number and cannot be smoothly deformed to the identity. These are referred to as 'large' gauge transformations. Thus the winding number is a characteristic of gauge transformations which takes discrete values and can be used to distinguish between classes of gauge transformations.

Since the winding number must be an integer the Chern–Simons action changes by an additive quantity of $2\pi k w$ under large gauge transformations, meaning that as long as the Chern–Simons level k is an integer the exponential term appearing in a path integral, e.g. the calculation of a vacuum expectation value for a spin network state Ψ,

$$\langle \Psi \rangle = \frac{\int \Psi e^{iS_{CS}} \mathcal{D}A}{\int e^{iS_{CS}} \mathcal{D}A} , \tag{7.13}$$

must be gauge-invariant, and hence so are the VEVs themselves.

The simplest observables we can calculate are Wilson loops.[7] A Wilson loop corresponds to a closed curve λ, which as already discussed can be used to probe the amount of curvature of a gauge field. Now consider a collection of r non-intersecting, closed curves, and endow each closed curve with an orientation, i.e. a preferred direction, which can be thought of as consistent set of vectors tangent to the curve at each point. The union of these closed curves will be referred to as a link, L. The closed curves in this link may interconnect with each other, and each closed curve may pass around and through itself forming a knot. There are a number of quantities that can be used to characterise links and knots, which we discuss in Appendix K.

The product of Wilson loops is referred to as a Wilson link, $W[L]$. Working in the Abelian case for simplicity of illustration, the expectation value of a Wilson link is (see [9])

$$\langle W(\lambda_1) \dots W(\lambda_r) \rangle = \langle W[L] \rangle = \exp \left(\frac{i2\pi}{k} \sum_{a,b=1}^{r} n_a n_b \Phi[\lambda_a, \lambda_b] \right) \tag{7.14}$$

where n_a, n_b are integers which we associate with irreps (see Sect. A.1) of U(1), and the Gauss linking number

$$\Phi[\lambda_a, \lambda_b] = \frac{1}{4\pi} \oint_{\lambda_a} dx_a^\mu \oint_{\lambda_b} dx_b^\nu \epsilon^{\mu\nu\rho} \frac{(x_a - x_b)^\rho}{|x_a - x_b|^3} \tag{7.15}$$

is an integer which counts the number of times one closed curve winds through the other. This linking number can be related to the flux linking associated with electromagnetism, if we think of one closed curve as a set of magnetic field lines, and the other as a wire loop which the field lines pass through. When $a = b$ we obtain the self-linking number $\Phi[\lambda_a, \lambda_a] = \Phi[\lambda]$, which diverges as we can see from Eq. (7.15). This divergence can be solved by creating a new closed curve, λ', derived from but also displaced from λ. To do this we "frame" λ, which is to say we assign a vector to every point of λ, and let the tips of these vectors be the points

---

[7] As we learned all the way back in Chap. 3 they are gauge-invariant, and we can construct other observables such as the field strength $F_{\mu\nu}$ from them.

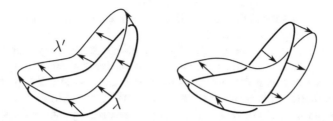

Fig. 7.3 A closed curve λ with framing, defining a second curve λ′. The two resultant curves can wind around each other, resulting in a non-zero linking number. If the "ribbon" between them is non-orientable then λ and λ′ are not actually distinct

defining λ′. In effect, we extrude λ from a one-dimensional closed curve into a ribbon with some finite width between its "left" and "right" sides (see Fig. 7.3).

We then treat these edges as separate knots. The problematic self-linking number $\Phi[\lambda]$ is then replaced by the linking number $\Phi[\lambda, \lambda']$. Of course, to ensure that λ and λ′ are truly distinct, the resulting ribbon defined between them must be an orientable surface, and hence their crossings must result in twists that are integer multiples of 2π.

This framing of curves, when applied to a spin network, turns each link into a ribbon and each node into a disk. The unframed (i.e. $\Lambda = 0$) spin networks could be analysed combinatorially using skein relations (see Appendix K) which embodied the representation theory of the group G determining the link labels. The extra framing in the $\Lambda \neq 0$ case depends on an extra parameter

$$q = \exp\left(\frac{\mathrm{i}2\pi}{k+h}\right) \tag{7.16}$$

where $h = 2$ in the SU(2) case. It should be noted that when we work in the spin-$\frac{1}{2}$ representation of SU(2) we find that the linking number of a Wilson link is the Kauffman bracket (Appendix K) evaluated with

$$A^4 = q = \exp\left(\frac{\mathrm{i}2\pi}{k+2}\right) \tag{7.17}$$

where in this case A is a specific parameter of the Kauffman bracket, not to be confused with A_μ.

We can view q as a deformation parameter, related to the framing of the spin networks. Now the relevant skein relations encode a labelling by representations of a q-deformed group $\mathrm{U}_q(\mathfrak{g})$ (also frequently called a "quantum group"). Despite the name, these are actually not groups, but rather $\mathrm{U}(\mathfrak{g})$ is the enveloping algebra[8] of the algebra $\mathfrak{g}$, and $\mathrm{U}_q(\mathfrak{g})$ is the enveloping algebra "deformed" by the parameter q.

---

[8] The basics of group theory and algebras are covered in Appendix A.1 but the reader in need of a more detailed discussion may consult the wikipedia article 'Universal enveloping algebra', Sect. 9.3 of [10], etc.

Such objects are a substantial subject in their own right, but we have attempted to address them briefly in Appendix L.

This construction achieves our stated goal of eliminating divergences. Writing BF theory in terms of Chern–Simons theory means we arrive at a version of spin networks and spin foams in which the labellings by representations of a group are replaced by representations of the corresponding q-deformed group. Similarly the labellings by intertwiners in terms of representations are replaced by intertwiners defined in terms of the q-deformed group's representations. In this case only representations of $U_q(\mathfrak{g})$ corresponding to $j = 0, \frac{1}{2}, \ldots \frac{k}{2}$ are permitted, placing a limit on the spin values associated to faces.

7.3   Some Recent Developments

Attempts to convert the conceptual view of spin foams, embodied in Eq. (7.1) into practical, tractable calculations require great ingenuity. Much effort has been expended on developing such calculations, a full discussion of which is beyond the scope of this book. But a common feature of such efforts centres upon choosing labelling schemes for the links and nodes in a spin network, and hence the associated faces, edges, and vertices of a spin foam. We have already seen in the case of BF theory and the development of loop quantum gravity that different choices of gauge group are possible, guided by the desire for computational ease and physical plausibility. In a similar fashion, labelling the elements of spin networks and spin foams by group representations, tensors, etc. has led to a number of interesting ideas and avenues for further research. We provide here a brief survey to whet the reader's appetite for their own investigations. We will start by discussing a very general concept, that of tensor networks, which hints at connections between quantum gravity, topology, and quantum computing, amongst other concepts, before reviewing some recent and more specific work by various authors.

7.3.1   Tensor Networks

We are all familiar with the idea of contracting tensor indices. In the case of a tensor acting on a vector, covector, or another tensor, we think of this as the action of a mapping from the space of objects being acted upon to construct a new mapping or quantity (e.g. the inner product). In many applications we can regard a series of operations as tensors themselves. The logic gates in a computer circuit, for example, behave like simple tensors, and composing several of these together can produce quite complex evaluations of the inputs provided to the circuit. Of course, a computer circuit can also be interpreted as a type of graph, with junctions and logic gates acting as hubs (to use the terminology we adopted in Sect. 6.1). It therefore becomes possible to view a series of tensors related by index contractions, a circuit, and a graph with appropriately labelled tracks and hubs as different versions of the same type of structure. That structure is what we mean when we refer to a tensor network.

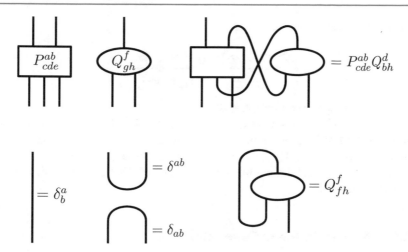

Fig. 7.4 Contraction of tensor indices using Penrose's graphical notation. On the top row we see two tensors (staggering of the indices has been suppressed to save space), and the product formed by two contractions between them. Different versions of the Kronecker delta, and a trace formed by contracting an upper and lower index on the same tensor are shown on the bottom row

Visualising interrelated tensors as forming a network becomes more intuitive when we employ a graphical notation due to Penrose [11]. In this notation a tensor with n raised indices and m lowered indices is drawn as a box or shape with n lines emerging from the top and m lines emerging from the bottom. Contractions between tensors are then represented by joining a line attached to the top of one box with a line attached to the bottom of another box (or possibly the same box), as illustrated in Fig. 7.4. Contraction of an upper index and a lower index reduces the number of 'free' lines in the diagram by exactly two. In this notation a scalar has no indices and hence no lines emerging from it, so if a tensor has one upper index (line) and one lower index (line) the contraction (joining) of the two gives a scalar—the trace of the tensor in question.

Various tensors of interest can be represented in this notation—for example the Kronecker delta is simply an unadorned line, either vertical, or with open ends pointing up or down (depending on whether it has mixed, raised, or lowered indices.) This is consistent with the way the Kronecker delta serves (in the summation convention) to equate indices, e.g. $T^{ab}\delta^c_a U^d_c = T^{ab}U^d_a$, and hence acts like (in the graphical notation) a line between the relevant tensors.

A symbol with a single upwards or downwards line stands in for a quantity with a single raised or lowered index—such as a vector with components v^a or a dual vector with components u_b. In the standard way we can think of a tensor with p raised indices and q lowered indices as being the tensor product of p vectors and q dual vectors, e.g. $T^{ab}{}_c = v^a \otimes w^b \otimes u_c$. If the vectors and dual vectors are elements of the representation space V and dual representation space V^* of some group G then we can view these composite tensors with p raised indices and q lowered indices as elements of the tensor product space

$$\underbrace{V^* \otimes \ldots \otimes V^*}_{q \text{ copies}} \otimes \underbrace{V \otimes \ldots \otimes V}_{p \text{ copies}} \,. \tag{7.18}$$

which is itself a representation space of G.

Further discussion of this notation can be found in [11, 12], and several informative reviews (especially for those with an interest in quantum information and quantum computing) can be found in [13, 14]. A very thorough discussion of the use of network diagrams to perform calculations involving coupling of spins, as discussed in Chap. 6, can be found in [15].

7.3.2   Spinorial LQG and Coherent States

We have at great length discussed the assignment of a representation of a group to the links in a spin network. We have also seen in Chap. 3 how holonomies correspond with the transport of spinors along a path. The spinorial formulation of LQG promotes the importance of this idea, labelling either end of each link in a spin network with a spinor in $\mathbb{C}^2$. This implies that the nodes are then labelled by several spinors, the number of spinors being equal to the valency of the node [16]. In this view the labelling by spinors is more fundamental, and any labelling of links by representations and nodes by intertwiners is derived from the spinors. Since the components of a spinor are complex numbers this approach has great promise to simplify calculations, by converting calculations involving representation theory into comparatively straightforward matters of complex analysis.

To begin, consider a spinor $|\zeta\rangle$ which takes values in $\mathbb{C}^2$, that is to say

$$|\zeta\rangle = \begin{pmatrix} \zeta_0 \\ \zeta_1 \end{pmatrix}, \qquad \langle\zeta| = \begin{pmatrix} \zeta_0^* & \zeta_1^* \end{pmatrix} \tag{7.19}$$

where the $*$ superscript denotes complex conjugation, and let the product of two spinors $|\zeta\rangle$ and $|\xi\rangle$ be taken in the usual way,

$$\langle\xi|\zeta\rangle = \xi_0^*\zeta_0 + \xi_1^*\zeta_1 \,. \tag{7.20}$$

Define a vector $\vec{X} = \langle\zeta|\vec{\sigma}|\zeta\rangle$. Then it is a fairly simple matter to confirm that

$$|\zeta\rangle\langle\zeta| = \frac{1}{2}\left(\langle\zeta|\zeta\rangle\,\mathbb{1} + \vec{X}\cdot\vec{\sigma}\right) \tag{7.21}$$

where the σ_a with $a \in \{1, 2, 3\}$ are the Pauli matrices. Furthermore $\vec{X}$ can be shown to be a null vector, i.e. $|\vec{X}|^2 = |X^0|^2$ where $X^0 = \langle\zeta|\zeta\rangle$. The vector $\vec{X}$ can be thought of as the normal vector to a face of a simplex, with magnitude proportional to the area of that face as discussed in Sect. 6.3.

We define a dual spinor, denoted $\widetilde{|\zeta\rangle}$, by the mapping[9]

$$\langle\zeta| \rightarrow \widetilde{|\zeta\rangle} = \begin{pmatrix} -\zeta_1^* \\ \zeta_0^* \end{pmatrix}. \tag{7.22}$$

Now given two spinors $|\zeta\rangle$ and $|\xi\rangle$ we claim that

$$g(\zeta,\,\xi) = \frac{|\zeta\rangle\widetilde{\langle\xi|} - \widetilde{|\zeta\rangle}\langle\xi|}{||\zeta||\,||\xi||} \tag{7.23}$$

is an element of SU(2). To confirm this we recognise that

$$g^\dagger(\zeta,\,\xi) = \frac{\widetilde{|\xi\rangle}\langle\zeta| - |\xi\rangle\widetilde{\langle\zeta|}}{||\zeta||\,||\xi||} \tag{7.24}$$

and easily check that $g^\dagger(\zeta,\,\xi)g(\zeta,\,\xi) = \mathbb{1}$, and also that $\det(g) = +1$.

Now when $g(\zeta,\,\xi)$ acts on normalised spinors we find that

$$g(\zeta,\,\xi)\frac{|\xi\rangle}{|\xi|} = -\frac{\widetilde{|\zeta\rangle}}{|\zeta|}, \qquad g^\dagger(\zeta,\,\xi)\frac{|\zeta\rangle}{|\zeta|} = \frac{\widetilde{|\xi\rangle}}{|\xi|}. \tag{7.25}$$

This is equivalent to the way we constructed the Schwinger line integral for parallel transport of a state vector along a path, $|\Psi'\rangle = U(g)|\Psi\rangle$.

The interpretation of $\vec{X}$ as the normal vector to a face implies a matching constraint. When two simplices are joined, pairing a face on one simplex with a face on the other, we expect the faces to have the same area. Using the idea that $|\zeta\rangle$ and $|\xi\rangle$ are the spinors on opposite ends of the link passing between (i.e. dual to) these two faces, and the relationship between components of a null vector, we can write the area matching constraint as

$$\mathbf{M} = \langle\zeta|\zeta\rangle - \langle\xi|\xi\rangle = 0. \tag{7.26}$$

We are used to constructing a Poisson bracket with the Hamiltonian and one "empty slot", forming an operator which describes the time evolution of functions. In a similar spirit we can create Poisson brackets using the area matching constraint, and find that

$$\{\mathbf{M},\,|\zeta\rangle\} = \mathbf{i}|\zeta\rangle \tag{7.27}$$

$$\{\mathbf{M},\,|\xi\rangle\} = -\mathbf{i}|\xi^*\rangle \tag{7.28}$$

from which we conclude that $\mathbf{M}$ acts upon the spinors to generate U(1) gauge transformations, $|\zeta\rangle \rightarrow e^{i\theta}|\zeta\rangle$.

---

[9] The notation $|\zeta]$ is more common for the dual spinor, but in the opinion of the author this notation is harder to read, since $|\zeta]$ looks almost indistinguishable from $[\zeta|$ and $|\zeta|$.

A further constraint implied by the normal vectors $\vec{X}$ arises from the fact that an n-valent node will have n spinors associated to it. This constraint, which may be written

$$\vec{c} := \sum_{i=1}^{n} \vec{X}_i = 0 \qquad (7.29)$$

can be interpreted as a requirement that the normal vectors sum to zero, or in other words the associated faces join together to form a closed simplex. Hence this is a closure constraint.

In [16–18] the labelling of n-valent nodes by spinors was used to investigate spatial simplices beyond the tetrahedra we have discussed so far. This generalisation carried over to polyhedra with n faces (with the restriction that exactly three faces met at a vertex).[10] In this polyhedral view of loop quantum gravity each spatial simplex is dual to a vertex labelled by an intertwiner carrying an irreducible representation of U(n).

It was already mentioned in Sect. 6.4 that tetrahedral simplices are not uniquely defined by the area of their faces. This admits an invariance under area-preserving diffeomorphisms of the simplices.

In the case of polyhedral simplices, it is possible to construct operators which increase and decrease surface area. These are essentially raising and lowering operators, exactly as we would expect in the case of harmonic oscillators, and are related to the interpretation of spin (angular momentum) labels determining the area of a face dual to a line carrying that label. Consider commutator relations for a pair of uncoupled harmonic oscillators,

$$[\hat{a},\, \hat{a}^\dagger] = 1 = [\hat{b},\, \hat{b}^\dagger], \qquad [\hat{a},\, \hat{b}] = 0. \qquad (7.30)$$

Using these, the generators of $\mathfrak{su}(2)$ are

$$\boldsymbol{J}^z \equiv \frac{1}{2}\left(\hat{a}^\dagger \hat{a} - \hat{b}^\dagger \hat{b}\right), \quad \boldsymbol{J}^+ \equiv \hat{a}^\dagger \hat{b}, \quad \boldsymbol{J}^- \equiv \hat{a}\hat{b}^\dagger. \qquad (7.31)$$

As one would expect from quantum field theory, the number operator is

$$\mathcal{E} = \frac{1}{2}\left(\hat{a}^\dagger \hat{a} + \hat{b}^\dagger \hat{b}\right) \qquad (7.32)$$

which we can interpret as a measure of the total energy or number of quanta (of area, in this case). It turns out that the Casimir

$$\boldsymbol{J}^2 = \mathcal{E}(\mathcal{E} + 1) \qquad (7.33)$$

---

[10] For instance, using the example of the Platonic solids, this admits the regular tetrahedron, cube, and dodecahedron, but not the octohedron or icosahedron.

and so indeed $\mathcal{E}$ can be regarded as a measure of area, as can be seen by comparison with Eq. (6.29). If we index the spins associated with different faces of a polyhedron by i and j then we have the operator

$$\left(\hat{a}_i^\dagger \hat{a}_j + \hat{b}_i^\dagger \hat{b}_j\right) \tag{7.34}$$

which raises spin j_i and lowers j_j by half a unit, thereby leaving total area invariant. The conjugate operator is defined by swapping $i \leftrightarrow j$. Furthermore, the operators

$$(\hat{a}_i \hat{b}_j - \hat{a}_j \hat{b}_i), \quad (\hat{a}_i \hat{b}_j - \hat{a}_j \hat{b}_i)^\dagger \tag{7.35}$$

respectively lower (by reducing j_i and j_j simultaneously by half a unit), and raise the total surface area of a polyhedron.

Consider an SU(2) gauge transformation applied to the n spinors labelling a particular n-valent node, $|\zeta_i\rangle \to |\zeta'_i\rangle$ for $i = 1, \ldots, n$. Then it is possible to define the SU(2)-invariant antisymmetric quantity

$$F_{ij} := \widetilde{\langle \zeta_i} | \zeta_j\rangle \tag{7.36}$$

as well as $E_{ij} := \langle \zeta_i | |\zeta_j\rangle$ which naturally is symmetric. The F_{ij} are used to define a creation operator in terms of spinor states,

$$F^\dagger = \frac{1}{2} \sum_{i \neq j} \widetilde{\langle \zeta_i} | \zeta_j\rangle (\hat{a}_i \hat{b}_j - \hat{a}_j \hat{b}_i)^\dagger . \tag{7.37}$$

The fact that we have operators that create and annihilate units of surface area of the polyhedra means we can think of a Fock space, with a vacuum state $|0\rangle$. The use of a rounded ket (or bra) indicates a coherent state. The creation operators in Eq. (7.37) can then act upon the vacuum to create coherent states $|\boldsymbol{J}, \zeta_i)$, where $\boldsymbol{J}$ is an integer equal to the area of the relevant face (which we denote in bold, consistent with the spin-area equivalence established in Chap. 6). The inner product of coherent states is found to be a straightforward power of the area, thus

$$(\boldsymbol{J}, \zeta_i | \boldsymbol{J}, \xi_i) = (\zeta_i | \xi_i)^{\boldsymbol{J}} . \tag{7.38}$$

Coherent states [19] are generally "well-behaved" under transformations, and can be thought of as states of minimal uncertainty, i.e. states for which the uncertainty principle yields an equality, rather than an inequality. One would hope then that they would shed light on the semi-classical limit of the theory. It turns out [16] that the coherent states are obtained by the action of U(N) on states corresponding to polyhedra (or nodes in the dual spin network in which all but two of the labelling spinors are zero.) To think of this process as creating arbitrary-valency nodes out of the vacuum makes intuitive sense, as an arrangement of two links meeting at such a bivalent node is equivalent to a single link with no node at all (i.e. the spin on one side of the "node" must equal the spin on the other side to conserve angular momentum,

Fig. 7.5 A Wilson loop
connecting a 2-valent vertex
to itself

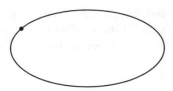

and so the associated intertwiner must simply be the identity.) The machinery thus developed makes it possible to evaluate expectation values of operators defined from Eq. (7.34). It is found that the mean values grow faster than the uncertainties as $J \to \infty$, showing that these coherent states peak as J becomes large.

Up until now we have been talking about individual spatial simplices. So let us step back to review some work [20] in which the spinorial approach to LQG is extend to cover the case of gluing simplices together to form spin networks, with a view to discussing spin foams. We begin by considering the simple case of a Wilson loop with both ends connecting to the same 2-valent node, as depicted in Fig. 7.5.[11]

Then as in Eq. (3.30) the loop holonomy is constructed and its trace taken to obtain

$$W_\lambda(g) = \frac{\widetilde{\langle \zeta_2 | \zeta_1 \rangle} + \langle \zeta_2 | \widetilde{\zeta_1} \rangle}{\|\zeta\| \, \|\xi\|} = \frac{F_{21} + F_{12}^*}{\sqrt{E_{11} E_{22}}}. \tag{7.39}$$

This is fine for a description of the classical phase space, requiring simplices to be closed and flat, and for any curvature in the spacetime they triangulate to occur at the edges between simplices, just as was described in the toy model of Sect. 6.1. The F_{ij} and their extension to the case of multi-node networks are classical quantities.

The quantisation of this picture associates a Hilbert space $\mathcal{H}_\Gamma$ to the spin network which is dual to this triangulation, as per Eq. (6.11). The question then arises as to whether $\mathcal{H}_\Gamma^{\text{spin}}$, the Hilbert space we would construct for a network with links labelled by pairs of spinors, corresponds with $\mathcal{H}_\Gamma$. In [20] a mapping was constructed between the Hilbert space of a single link labelled by a spin, and a single link labelled by pairs of spinors. This mapping was then extended to networks with an arbitrary number of nodes and links.

Let $\mathcal{U}$ be the space of holomorphic[12] square-integrable functions over the complex numbers, $L_{\text{hol}}^2(\mathbb{C}, d\mu)$, where for this discussion we don't need to consider the form of the measure $d\mu$ (see [20] for more details). This is a space of polynomials in powers of $z \in \mathbb{C}$, and so it has an orthonormal basis given by $e_n(z) = z^n/\sqrt{n!}$. More generally we would write $\mathcal{U}_n = L_{\text{hol}}^2(\mathbb{C}^n, d\mu)$. The spinors we have considered so far transform under SU(2), and thus we are led to consider $\mathcal{U}_2$. This space has as an orthonormal basis

$$e_m^j(z) = \frac{1}{\sqrt{(j+m)!(j-m)!}}(z_0)^{j+m}(z_1)^{j-m} \equiv \frac{1}{\sqrt{x!y!}}(z_0)^x(z_1)^y \tag{7.40}$$

---

[11] This looks like a bit of a cheat, as it's equivalent to the node-less link we mentioned above in the case of the Fock vacuum, but still serves to introduce a simple version of some results which will be described shortly.

[12] I.e. They can be expanded in powers of a complex number z, and don't depend on z^*.

where we now invoke two complex numbers z_0, z_1 corresponding to the two complex components of a spinor. Equation (7.40) illustrates the connection between familiar notation in terms of total and magnetic quantum numbers (with j taking half-integer values and m taking values between $\pm j$) and points (x, y) where x, y are natural numbers, defining the form of a polynomial in powers of two complex numbers.

This allows the construction of the Hilbert space for a link labelled by a spinor at either end. It is given by two copies of the space of functions for a spinor, however we must take into account the matching constraint which as we have seen generates U(1) gauge transformations. Hence the Hilbert space for a link is identified as

$$\mathcal{H}_l^{\text{spin}} = (\mathcal{U}_2 \times \mathcal{U}_2)\,/\mathrm{U}(1) \tag{7.41}$$

with orthonormal basis states $e_{m_1}^j \otimes e_{m_2}^j$ (one might expect the occurrence of two different values, j_1 and j_2, however it can be shown that all states with $j_1 \neq j_2$ vanish).

The Hilbert space of an individual link does not pay attention to the SU(2) gauge transformations that may be applied to the n spinors labelling an n-valent node. The transition to considering the Hilbert space of entire networks labelled by spinors can be achieved by constructing SU(2) invariant functions corresponding to nodes, and connect these in such a way as to fulfill the U(1) invariance required by the matching constraint on each link between nodes. It turns out that, for n-valent nodes, polynomials in the variables F_{ij} defined in Eq. (7.36) are elements of $\mathcal{U}_2^{\otimes n}$, and hence can be used (nodewise) to construct functions of the spin networks, extending the result of Eq. (7.39). Specific details can be found in [20], but the basic result is that functions of the F_{ij} are much easier to work with than expressions involving SU(2) recoupling theory applied to the intertwiners labelling spin network nodes.

7.3.3   Group Field Theory

Up to this point our ideas of spin networks, and spin foams, have had a fairly simple form which (if technically challenging) is at least intuitively simple; choose a base manifold which we can think of as similar if not identical to $\mathbb{R}^N$, embed a graph in that manifold, label the tracks and hubs of the graph, and ultimately consider only the labelled graph so that the base manifold becomes superfluous. In group field theory [21] the base manifold is instead taken to be the group manifold of a Lie group. This means that points in the manifold are no longer explicitly positions in some version of space (or spacetime), but group elements defined by particular choices of parameters.

To explore this idea, let us define a field ϕ which is a mapping from d copies of a Lie group manifold to the complex numbers

$$\phi : G^d \to \mathbb{C} \tag{7.42}$$

where d is the number of spacetime dimensions we wish to model. Given d group elements $\{g_i\} \in G^d$ we have

$$\phi(g_1, \ldots, g_d) \to \int \left(\prod_{i=1}^{d} dg_i{}' \right) U(g_1', g_1) \times \ldots \times U(g_d', g_d)\phi(g_1', \ldots, g_d')$$

(7.43)

where the $U(g_i', g_i)$ are elements of some unitary group U, and are analogous to the Schwinger line integral terms which we encountered in Eq. (3.26). These terms were identified as gauge rotations, and so in this case the U are mappings between $\{g_i\}$ and $\{g_i'\}$. Consequently we can view ϕ as a tensor transforming under d copies of the group U, where the $\{g_i\}$ are the tensor index labels.

We can define an action for our group field theory consisting of kinetic and interaction terms,

$$S_d[\phi] = S_{\text{kin}} + S_{\text{int}}$$

(7.44)

where the kinetic term takes the form

$$S_{\text{kin}} = \frac{1}{2} \int dg_i dg_i' \phi(g_i) K(g_i g_i'^{-1})\phi(g_i')$$

(7.45)

and the kinetic kernel K is invariant for all $h, h' \in G$,

$$K(hg_i g_i'^{-1} h') = K(g_i g_i'^{-1}).$$

(7.46)

To represent the couplings inherent to the interaction term we require $i \neq j$ and write

$$\phi(g_{1j}) = \phi(g_{12}, g_{13}, \ldots g_{1(d+1)})$$
$$\phi(g_{2j}) = \phi(g_{21}, g_{23}, \ldots g_{2(d+1)})$$

(7.47)

and so forth, where the physical interpretation of g_{ij} will be developed below. The interaction term is then

$$S_{\text{int}} = \frac{\mu}{d+1} \int \prod_{i=1}^{d+1} \prod_{j(\neq i)=1}^{d+1} dg_{ij} V(g_{ij} g_{ji}'^{-1})\phi(g_{ij})$$

(7.48)

where dg_{ij} is an invariant (Haar) measure on G, and μ is a coupling strength. If it is understood that the measure includes a product over i values, so that $dg_{ij} \equiv dg_{1j} \ldots dg_{(d+1)j}$ then we can write the interaction term as

$$S_{\text{int}} = \frac{\mu}{d+1} \int \prod_{j(\neq i)=1}^{d+1} dg_{ij} V(g_{ij} g_{ji}^{-1})\phi(g_{1j}) \ldots \phi(g_{(d+1)j}).$$

(7.49)

As with K, the interaction kernel V is an invariant, where for all $h_i \in G$,

$$V(h_i g_{ij} g_{ji}'^{-1} h_j^{-1}) = V(g_{ij} g_{ji}'^{-1}).$$

(7.50)

For the sake of concreteness, we will look at an example in $d = 4$ dimensions. We draw inspiration from the Barrett–Crane model (see Sect. 7.2), and choose the gauge field to be SO(4). The ϕ then become a field

$$\phi(g_1, g_2, g_3, g_4) : \text{SO}(4) \times \text{SO}(4) \times \text{SO}(4) \times \text{SO}(4) \rightarrow \mathbb{R} . \tag{7.51}$$

The action is found fairly easily from Eqs. (7.45) and (7.49). A simple choice for the kinetic kernel is

$$K(g_i g_i'^{-1}) = \int_G dh \prod_{i=1}^{4} \delta(g_i g_i'^{-1} h) . \tag{7.52}$$

In this case (glossing over symmetries and constraints) the kinetic part of the action is

$$S_{\text{kin}} = \frac{1}{2} \int \prod_{i=1}^{4} dg_i \phi^2(g_1, g_2, g_3, g_4) . \tag{7.53}$$

To make the interaction term easier to write, let us reduce somewhat the number of indices present by defining $\phi_A = \phi(g_{1j})$ and similarly $\phi_B = \phi(g_{2j})$ etc. as per Eq. (7.47). In full then, we have

$$\phi_A = \phi(g_{12}, g_{13}, g_{14}, g_{15}) \tag{7.54a}$$
$$\phi_B = \phi(g_{21}, g_{23}, g_{24}, g_{25}) \tag{7.54b}$$
$$\phi_C = \phi(g_{31}, g_{32}, g_{34}, g_{35}) \tag{7.54c}$$
$$\phi_D = \phi(g_{41}, g_{42}, g_{43}, g_{45}) \tag{7.54d}$$
$$\phi_E = \phi(g_{51}, g_{52}, g_{53}, g_{54}) . \tag{7.54e}$$

Remembering the implicit iteration through values of i, the potential is then

$$S_{\text{pot}} = \frac{\mu}{5} \int \prod_{j(\neq i)=1}^{5} dg_{ij} \phi_A \phi_B \phi_C \phi_D \phi_E . \tag{7.55}$$

We can give a physical meaning to the integrand of Eq. (7.53) by reference to a series of graph diagrams [6]. We will use the hub-track terminology for abstract graphs, but very shortly see how these diagrams relate to simplices and spin networks. To start, represent ϕ as a tetravalent hub. Note that in general we will consider several hubs labelled by i, explaining the notation g_{ij} used above, as referring to track j attached to hub i. As we are only considering one hub (for the moment) we simplify the indices accordingly. We therefore label the four tracks of the hub $g_1, \ldots g_4$, as in Fig. 7.6, and write $\phi(g_j) = \phi(g_1, g_2, g_3, g_4)$. A consequence of the simplification of indices is that the graph diagrams that will now be introduced are also simplified. This will be rectified shortly.

For the moment, recall the discussion in Sect. 7.3.1, in which we consider the joining of lines to be equivalent to contraction of indices. Setting the indices of two

Fig. 7.6 Interpretation of ϕ as a hub attached to four tracks, corresponding to four index labels

Fig. 7.7 Contraction of two ϕ tensors, corresponding to the connection of two hubs via their four tracks

Fig. 7.8 Contraction of five hubs leads to the identification of ϕ_A, ϕ_B, ϕ_C, ϕ_D, ϕ_E as vertices of a 4-dimensional simplex. The correspondence to the 4-simplices in Fig. 7.2a should be apparent

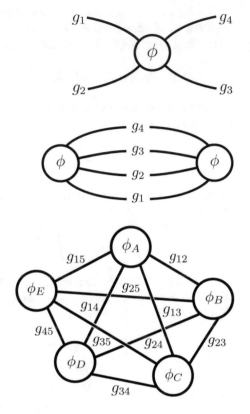

hubs equal we obtain ϕ^2, exactly as we are used to doing with tensor indices to obtain, for instance, $F^{\mu\nu}F_{\mu\nu} = F^2$. Diagrammatically we can view this as taking two hubs and joining their four tracks as Fig. 7.7 depicts.

The potential part of the action involves ϕ_A, ϕ_B, ϕ_C, ϕ_D, ϕ_E, which we interpret as the contraction of five hubs with each other. Since there are five hubs with four tracks each we have twenty indices, but each index is common to two hubs, so there are in fact only ten unique sets of indices. One will therefore sometimes encounter the potential term written in the form

$$S_{\text{pot}} = \frac{\mu}{5!} \int \prod_{k=1}^{10} dg_k \phi_A \phi_B \phi_C \phi_D \phi_E \qquad (7.56)$$

with k (not to be confused with the Chern–Simons level of Sect. 7.2.2) standing in for pairs of i, j values. These values of k correspond to the ten edges of a 4-simplex, with the five hubs corresponding to its vertices, as described in Sect. 7.1. Performing the contractions between ϕ_A, ϕ_B, ϕ_C, ϕ_D, ϕ_E we obtain a diagram as in Fig. 7.8. Examining the resultant connections and identifying g_{ij} as the group element which labels the edge connecting vertex i with vertex j, we see that the indices associated to each hub are as given in Eqs. (7.54a)–(7.54e).

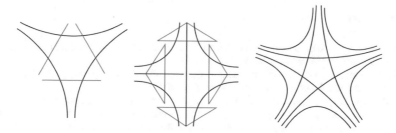

Fig. 7.9 Contraction of ϕ tensors in two (left), three (middle), and four dimensions. In the two-dimensional case each ϕ has two lines running through it, and corresponds to a linear face—in blue—of a 2-simplex (a triangle). In three dimensions has three lines run through each ϕ, corresponding to a triangular face of a 3-simplex (a tetrahedron). The tetrahedral faces of the 4-simplex (right) are omitted for clarity, but a quick comparison shows that Fig. 7.8 is the same diagram with "external" lines omitted

Now it is tempting to identify the five hubs in Eqs. (7.54a)–(7.54e) as vertices of a volume simplex in four dimensions. We can understand the potential term Eq. (7.53) as being based on the structure of such simplices. However, we should actually recognise that the ϕ can be associated with four *lines* (and so four of the tracks per hub have been suppressed so far, just as we suppressed one index on g_{ij} above.) So we should think of each hub as a tetrahedral face of a 4-simplex, and think of g_{ij} as representing the line running through faces i and j. This conception is illustrated in Fig. 7.9, in two, three, and four dimensions. Likewise, the coupling of two ϕ factors in the kinetic part of the action, as illustrated in Fig. 7.7, should be thought of as a set of four lines passing into and through one ϕ, through the next ϕ and continuing out the other side.

This leaves us with a scheme that can very easily be viewed as a Feynman diagram for interactions between an initial spin network state and a final state. We can therefore conceive of performing a perturbative sum over all topologies and geometries of the simplices. In this way we connect up to the concept of a path integral over field configurations involving a perturbative sum of interaction terms described by Feynman diagrams, as outlined in Sect. 3.4. This in turn should enable the calculation of the transition amplitudes mentioned above. For further details, the reader is recommended to consult the reviews [22,23]. Other interesting reviews of the spin foam formalism and group field theory include [24,25].

The time has come to move on from our survey of approaches to the quantum dynamics of spacetime. In each case the goal has been to construct something recognisable as a path integral formulation of the structure of spacetime, in other words, the sum-over-histories for our quantum version of general relativity. Hopefully this discussion has set the stage for the reader to explore these concepts on their own. We now turn to looking at some applications of the ideas we have covered so far.

References

1. J.C. Baez, Spin foam models. Class. Quant. Grav. **15**, 1827–1858 (1998). https://doi.org/10.1088/0264-9381/15/7/004. arXiv:gr-qc/9709052
2. M. Reisenberger, Worldsheet formulations of gauge theories and gravity (1994). https://doi.org/10.48550/arXiv.gr-qc/9412035. arXiv: gr-qc/9412035
3. M.P. Reisenberger, C. Rovelli, "Sum over surfaces" form of loop quantum gravity. Phys. Rev. D **56**, 3490–3508 (1997). https://doi.org/10.1103/PhysRevD.56.3490. arXiv:gr-qc/9612035
4. J.C. Baez, An introduction to spin foam models of BF theory and quantum gravity. Lect. Notes Phys. **543**, 25–94 (2000). https://doi.org/10.48550/arXiv.gr-qc/9905087. arXiv:gr-qc/9905087
5. J.W. Barrett, L. Crane, Relativistic spin networks and quantum gravity. J. Math. Phys. **39**, 3296–3302 (1998). https://doi.org/10.1063/1.532254. arXiv: gr-qc/9709028
6. R. De Pietri et al., Barrett-Crane model from a Boulatov-Ooguri field theory over a homogeneous space. Nucl. Phys. B **574**, 785 (2000). https://doi.org/10.1016/S0550-3213(00)00005-5. arXiv:hep-th/9907154
7. E. Alesci, C. Rovelli, The complete LQG propagator: I. Difficulties with the Barrett-Crane vertex. Phys. Rev. D **76**, 104012 (2007). https://doi.org/10.1103/PhysRevD.76.104012. arXiv:0708.0883v1
8. J.C. Baez, J.P. Muniain, *Gauge Fields, Knots, and Gravity (Series on Knots and Everything)*, vol. 4 (World Scientific Pub Co Inc, 1994). ISBN: 9810220340. https://doi.org/10.1142/2324
9. D. Grabovsky, *Chern–Simons Theory in a Knotshell* (2022). https://web.physics.ucsb.edu/davidgrabovsky/filesnotes/CSandKnots.pdf
10. B.C. Hall, *Lie Groups, Lie Algebras, and Representations An Elementary Introduction*, 2nd edn. (Springer, 2015). ISBN: 978-3-319-13466-6. https://doi.org/10.1007/978-3-319-13467-3
11. R. Penrose, Applications of negative dimensional tensors. Comb. Math. Appl. 221–244 (1971). Ed. by D. Welsh
12. B. Coecke, A. Kissinger, *Picturing Quantum Processes* (Cambridge University Press, 2017). ISBN: 9781107104228. https://doi.org/10.1017/9781316219317
13. J.C. Bridgeman, C.T. Chubb, Hand-waving and interpretive dance: an introductory course on tensor networks. J. Phys. A: Math. Theor. **50**, 223001 (2017). https://doi.org/10.1088/1751-8121/aa6dc3. https://iopscience.iop.org/article/10.1088/1751-8121/aa6dc3/pdf
14. (Various authors). Review Articles and Learning Resources. https://tensornetwork.org/reviews_resources.html
15. V. Aquilanti et al., Semiclassical mechanics of the wigner 6j-symbol. J. Phys. A: Math. Theor. **45**, 065209 (2012). https://doi.org/10.1088/1751-8113/45/6/065209. arXiv:1009.2811v2
16. L. Freidel, E.R. Levine, U(N) coherent states for loop quantum gravity. J. Math. Phys. **52**, 052502 (2011). https://doi.org/10.1063/1.3587121. arXiv:1005.2090v1
17. F. Girelli, E.R. Livine, Reconstructing quantum geometry from quantum information: spin networks as harmonic oscillators. Class. Quant. Grav. **22**, 3295–3314 (2005). https://doi.org/10.1088/0264-9381/22/16/011. arXiv: gr-qc/0501075
18. L. Freidel, E.R. Levine, The fine structure of SU(2) intertwiners from U(N) representations. J. Math. Phys. **51**, 082502 (2010). https://doi.org/10.1063/1.3473786. arXiv: 0911.3553
19. A.M. Perelomov, Commun. Math. Phys. **26**, 222 (1972). https://doi.org/10.1007/bf01645091. arXiv:math-ph/0203002
20. E.R. Levine, J. Tambornino, Spinor representation for loop quantum gravity. J. Math. Phys. **53** (2012). https://doi.org/10.1063/1.3675465. arXiv:1105.3385v2
21. D.V. Boulatov, A model of three-dimensional lattice gravity. Mod. Phys. Lett. A **7**, 1629 (1992). https://doi.org/10.1142/S0217732392001324. arXiv:hep-th/9202074
22. L. Freidel, Group field theory: an overview. Int. J. Theor. Phys. **44**, 1769–1783 (2005). https://doi.org/10.1007/s10773-005-8894-1. arXiv:hep-th/0505016v1

23. D. Oriti, Group field theory and loop quantum gravity. 100 Years of General Relativity **4**, 125–151 (2017). https://doi.org/10.48550/arXiv.1408.7112. arXiv:1408.7112v1
24. A. Perez, The spin-foam approach to quantum gravity. Living Rev. Relativ. **16**, 3 (2013). https://doi.org/10.12942/lrr-2013-3. arXiv:1205.2019
25. S. Gielen, L. Sindoni, Quantum cosmology from group field theory condensates: a review. SIGMA **12**, 082 (2016). https://doi.org/10.3842/SIGMA.2016.82. arXiv: 1602.08104v2

Applications

8

Ultimately, the value of any theory is judged by its relevance for the *real* world. Unfortunately, due to the small length scales involved, direct tests of models of quantum gravity are not easy to perform. However one can try to reproduce well-known results from other physical theories as a preliminary consistency test for newer theories. In this section, we will consider how LQG can be applied to the calculation of black hole entropy, and cosmological models.

While the question of black hole entropy is, as yet, an abstract problem, it is concrete enough to serve as a test-bed for theories of quantum gravity. In addition to the Bekenstein area law (mentioned in Chap. 1), by investigating the behavior of a scalar field in the curved background geometry near a black hole horizon it was determined [1] that all black holes behave as almost perfect black bodies radiating at a temperature inversely proportional to the mass of the black hole, $T \propto 1/M_{BH}$. This thermal flux is named Hawking radiation after its discoverer. These properties of a black hole turn out to be completely independent of the nature and constitution of the matter which underwent gravitational collapse to form the black hole in the first place. These developments led to the understanding that a macroscopic black hole, at equilibrium, can be described as a thermal system characterized solely by its mass, charge and angular momentum.

Bekenstein's result has a deep implications for any theory of quantum gravity. The "Bekenstein bound" refers to the fact that Eq. (1.1) is the *maximum* number of degrees of freedom—of both, geometry and matter–that can lie within *any region of spacetime* of a given volume V. The argument is straightforward [2]. Consider a region of volume V whose entropy is greater than that of a black hole which would fit inside the given volume. If we add additional matter to the volume, we will eventually trigger gravitational collapse leading to the formation of a black hole, whose entropy will be less than the entropy of the region was initially. However, such a process would violate the second law of thermodynamics and therefore the entropy of a given volume must be at a maximum when that volume is occupied by a black hole. And since the entropy of a black hole is contained entirely on its

© Springer Nature Switzerland AG 2024
S. Bilson-Thompson, *Loop Quantum Gravity for the Bewildered*,
https://doi.org/10.1007/978-3-031-43452-5_8

horizon, one must conclude that the maximum number of degrees of freedom N_{max} that would be required to describe the physics in a given region of spacetime $\mathcal{M}$, in any theory of quantum gravity, scales not as the volume of the region $V(\mathcal{M})$, but as the area of its boundary [2,3] $N_{max} \propto A(\partial\mathcal{M})$.

In view of the independence of the Bekenstein entropy on the matter content of the black hole, the origin of Eq. (1.1) must be sought in the properties of the horizon geometry. Assuming that at the Planck scale, geometrical observables such as area are quantized such that there is a minimum possible area element a_0 that the black hole horizon, or any surface for that matter, can be "cut up into", Eq. (1.1) can be seen as arising from the number of ways that one can put (or "sew") together N quanta of area to form a horizon of area $A = kNa_0$, where k is a constant. In this manner, understanding the thermal properties of a black hole leads us to profound conclusions:

1. In a theory of quantum gravity the physics within a given volume of spacetime $\mathcal{M}$ is completely determined by the values of fields on the boundary of that region $\partial\mathcal{M}$. This is the statement of the *holographic principle*.
2. At the Planck scale (or at whichever scale quantum gravitational effects become relevant) spacetime ceases to be a smooth and continuous entity, i.e. *geometric observables are quantized*.

In LQG, the second feature arises naturally—though not all theorists are convinced that geometry should be "quantized" or that LQG is the right way to do so. One can also argue on general grounds that the first feature—holography—is also present in LQG, though this has not been demonstrated in a conclusive manner. Perhaps this book might motivate some of its readers to close this gap!

Let us now review the black hole entropy calculation in the framework of LQG.

8.1   Black Hole Entropy

The ideas of *quantum geometry* allow us to give a statistical mechanical description of a black hole horizon. This is analogous to the statistical mechanical description of entropy for a gas, or some other system composed of many smaller parts, and is related to the concept, mentioned above, that the horizon of a black hole can be cut up into small area elements. Just as a gas in classical thermodynamics can have a macrostate defined by its pressure, temperature, etc. and microstates defined by the positions and momenta of its constituent molecules, we are led to the idea that a black hole can have macrostates defined by mass, charge, and angular momentum, and microstates defined by the properties of the area elements its boundary has been subdivided into.

In classical thermodynamics, the entropy of a system, S, was related to the number of microstates, Ω, by Boltzmann by the formula

$$S = k_B \ln \Omega \tag{8.1}$$

where k_B is Boltzmann's constant. Gibbs deduced a similar formula relating the entropy to the probability, p_i, of a given microstate occurring,

$$S = -k_B \sum_i p_i \ln p_i \ .$$ (8.2)

For a proof that the logarithmic definitions of entropy provided by Boltzmann and Gibbs correspond to the thermodynamic definition usually encountered first in undergraduate courses,

$$\mathrm{d}S \geq \frac{\mathrm{d}Q}{T}$$ (8.3)

the reader is referred to [4].

It is perhaps worth taking a small digression to discuss these microstate-based equations in more detail, since they suggest that the entropy of a black hole can indeed be calculated by reference to the microstates of quantum geometry at its boundary. Many discussions of entropy quote the result that it is a logarithmic function of the number of microstates, Ω but a clear-cut explanation of why can be hard to come by. Indeed the formula

$$S = k_B \ln \Omega$$

is often "dropped in" to discussions and derivations alike, and explained (if at all) by saying that it turns a large number (the number of microstates) into a smaller, more manageable number, and that it has the correct properties to describe the disorder of a physical system.[1] This may seem half-hearted and unsatisfying to many readers (it certainly does to the author), so in Appendix M a discussion is provided which is, if not entirely devoid of hand-waving, at least somewhat less vague. This discussion is largely based on Shannon's work on entropy in the context of information theory [5].

In general there are two ways to calculate the entropy associated with a given random variable x:

1. Using Shannon's formula. Let us say that we sample our random variable from some given ensemble, from which we draw N samples. The variable x takes values in the set $\{x_i\}$ where $i = 1, 2, \ldots n$. Then the entropy associated with our lack of knowledge of the variable x is given by

$$S(x) = -\sum_{i=1}^{n} p(x_i) \ln p(x_i)$$ (8.4)

---

[1] For instance if you combine two systems with entropy S_1 and S_2, the resulting system has entropy $S = S_1 + S_2$. This is not an entirely trivial or universal property, since if you combine two systems at temperatures T_1 and T_2, the resulting system does not attain a temperature $T = T_1 + T_2$.

where $p(x_i)$ is the probability that the random variable takes on the value x_i. If in the N samples on which the entropy is based, the ith value x_i occurs k_i times (with the constraint that $\sum_i k_i = N$), then we have the usual frequentist definition for the probability associated with that value,

$$p(x_i) = \frac{k_i}{N}.$$

The definition of the Shannon entropy (8.4) is equivalent to the definition of the Gibbs entropy in statistical mechanics.

2. Using the statistical mechanics method, or its more general version, Jaynes' formalism [4]. This is based on the *maximum entropy principle*, according to which, in the absence of any prior information about a given random variable the *least unbiased* assumption one can make is that the variable satisfies a probability distribution which possesses the maximum possible entropy. This assumption leads us to the usual Boltzmann form of the probability. For a given value of the random variable x_i, the associated probability distribution must satisfy the maximum entropy criterion (wherein (8.4) is maximized) and also the usual axioms of probability theory

$$\sum_{i=1}^{n} p_i = 1, \tag{8.5a}$$

$$\langle f(x) \rangle = \sum_{i=1}^{n} p_i f(x_i) \tag{8.5b}$$

where $f(x)$ is any function of x. The unique probability function which satisfies these criteria is found to be (see for e.g. [6, Sect. 3.2])

$$p_i = e^{-\alpha - \beta x_i} \tag{8.6}$$

where α, β are Lagrange multipliers required for enforcing the constraints given in (8.5)[2] and where α, β can be identified with the chemical potential and inverse temperature respectively, associated with the random variable x. Using (8.6) we can write down the partition function

$$Z(\beta) = \sum_{i=1}^{n} e^{-\alpha - \beta x_i} \tag{8.7}$$

---

[2] The quantity being extremised has the form

$$L = -\sum_{i=1}^{n} \{ p(x_i) \ln p(x_i) - \alpha p(x_i) - \beta f(x_i) p(x_i) \}.$$

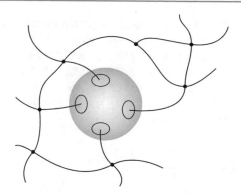

Fig. 8.1 A spin network corresponding to some state of geometry in the bulk punctures a black hole horizon at the indicated locations. Each puncture yields a quantum of area (depicted by the black ovals) proportional to $\sqrt{j(j+1)}$ where j is the spin-label on the corresponding edge. The entropy of the black hole—or, more precisely, of the horizon—can be calculated by counting the number of possible configurations of punctures which add up to give a macroscopic value of the area lying within some finite interval $(A, A + \delta A)$

given which we can evaluate the usual thermodynamic quantities such as expectation values, free energy and the entropy in x, given by

$$\langle f(x) \rangle = -\frac{\partial \ln Z(\beta)}{\partial \beta} \tag{8.8a}$$

$$F(T) = -kT \ln Z(T) \tag{8.8b}$$

$$S = -\frac{\partial F}{\partial T} \tag{8.8c}$$

where the inverse "temperature" is given by $\beta = 1/kT$.

In the case of quantum geometry, the microstates correspond with the assignments of area to the discrete "pieces" of a surface (such as the event horizon of a black hole). Hence for each macroscopic interval of area in the range $[A + \delta A, A - \delta A]$, entropy S is proportional to the log of the number of ways in which we can puncture the sphere to yield an area within that interval (Fig. 8.1).

The state of a quantum surface is specified by a sequence of N integers (or half-integers depending on the gauge group) $\{j_i, \ldots, j_N\}$, each of which labels an edge which punctures the given surface. The area of the surface is given by a sum over the Casimir at each puncture,

$$\mathbf{A} = 8\pi\gamma l_P^2 \sum_{i=1}^{N} \sqrt{j_i(j_i + 1)}. \tag{8.9}$$

The eigenvalues of the operator j_i are of the form $k_i/2$, where $k_i \in \mathbb{Z}$. Thus, the eigenvalues of the area operator are of the form

$$A_i = 4\pi\gamma l_P^2 \sqrt{k_i(k_i + 2)} = 4\pi\gamma l_P^2 \sqrt{(k_i + 1)^2 - 1}. \tag{8.10}$$

In addition to (8.10) the integers $\{k_I\}$ must also satisfy a so-called *projection constraint*, which is discussed later in this section.

The task at hand is the following; given an interval $[A + \delta A, A - \delta A]$, where A is a macroscopic area value and δA is some small interval ($\delta A/A \ll 1$), and the number N of edges which puncture the surface, determine the allowed the number $N(M)$ of sequences of integers $\{k_i, \ldots, k_N\}$, such that the resulting value for the total area falls within the given interval

$$M = \frac{A}{4\pi\gamma l_P^2} = \sum_i \sqrt{k_i(k_i + 2)} \in [A + \delta A, A - \delta A]. \tag{8.11}$$

There are various approaches to this problem. We summarize two of these—the simple argument of Rovelli's [7] and the number theoretical approach of [8,9] in the next section, and in the following section which we describe the approach based on Chern-Simons theory with SU(2) gauge group.

8.1.1 Rovelli's Counting

We want to compute the number of sequences $N(M)$, where each sequence $\{k_i\}$ satisfies

$$M = \frac{A}{4\pi\gamma l_P^2} = \sum_i \sqrt{k_i(k_i + 2)}.$$

Let us first note the following set of inequalities:

$$\sum_i \sqrt{k_i^2} < \sum_i \sqrt{k_i(k_i + 2)} \equiv \sum_i \sqrt{(k_i + 1)^2 - 1} < \sum_i \sqrt{(k_i + 1)^2}. \tag{8.12}$$

Let $N_+(M)$ denote the number of sequences such that $\sum_i k_i = M$ and $N_-(M)$ denote the number of sequences such that $\sum_i (k_i + 1) = M$). Then the above set of inequalities implies that [7]

$$N_-(M) < N(M) < N_+(M). \tag{8.13}$$

Computing $N_+(M)$ boils down to counting the number of partitions of M, i.e. the numbers of sets of *ordered*, positive integers whose sum is M. As noted in [7], this can be solved by observing that if $(k_1, k_2, \ldots, k_n)$ is a partition of M, then $(k_1, k_2, \ldots, k_n, 1)$ and $(k_1, k_2, \ldots, k_n + 1)$ are partitions of $M + 1$. All partitions of $M + 1$ can be obtained in this manner and therefore we have $N_+(M + 1) = 2N_+(M)$, which implies that $N_+(M) = C2^M$, where C is a constant.

8.1.2   Number Theoretical Approach

This approach consists of two steps:

A.   Determine allowed sequences. This involves solving the Brahmagupta-Pell (BP) equation.[4] For now, we will work in units where $4\pi\gamma l_P^2 \equiv 1$. Thus for a given set of N punctures on a quantum horizon, the total area can be written as

$$A = \sum_{i=1}^{N} A_i = \sum_{i=1}^{N} \sqrt{(k_i + 1)^2 - 1}\,.$$

For each possible value of k, let g_k be the number of punctures which have the corresponding eigenvalue. So, we can write

$$A = \sum_{k=1}^{k_{\max}} g_k \sqrt{(k + 1)^2 - 1}$$

with $g_k = 0$ if no puncture has spin $k/2$. Clearly the sum over all possible values of k gives the total number of punctures on the horizon, $\sum_k g_k = N$. As shown in Appendix N, the square root of any integer can be written as the product of an integer and the square-root of a square-free integer. Since $k \in \mathbb{Z} \Rightarrow (k + 1)^2 - 1 \in \mathbb{Z}$, therefore we can write

$$\sqrt{(k + 1)^2 - 1} = y_k \sqrt{p_k}$$

for some $y_k \in \mathbb{Z}$ and $p_k \in \mathbb{A}$, where $\mathbb{A}$ is the set of square-free integers. This implies that the area eigenvalue can be written as an integer linear combination of square-roots of square-free numbers,

$$A = \sum_{i=1}^{i_{\max}} y_i \sqrt{p_i}\,,$$

---

[4] It is well-known that the name of "Pell's Equation" was the result of Leonhard Euler's misidentification of John Pell with the mathematician Lord Brouckner. If we gave Euler a second chance to name the equation, he might have called it "Brouckner's equation". This equation had previously been intensively studied by the Indian mathematicians Brahmagupta and Bhaskara around the 5th century B.C. and 12th century A.D. respectively. However, Brouckner and Euler are to be forgiven for not having knowledge of the existence of this earlier work. The authors hereby take the liberty of correcting this historical wrong associated with the naming of this equation, by adding the prefix "Brahmagupta" to the presently accepted name "Pell's Equation".

leading us to the condition that

$$\sum_{k=1}^{k_{max}} g_k \sqrt{(k+1)^2 - 1} = \sum_{i=1}^{i_{max}} y_i \sqrt{p_i} \,.$$

As a first step towards solving the general case, let us first try to determine the solution of the above equation for a single area eigenvalue $k_i/2$,

$$\sqrt{(k_i + 1)^2 - 1} = y_i \sqrt{p_i}$$

knowing which we will be able to solve the general equation. Here the unknown variables are k_i, y_i. The p_i are the known square-free numbers. Setting $x_i = k_i + 1$ and squaring both sides we obtain

$$x_i^2 - p_i y_i^2 = 1.$$

This is commonly known as Pell's equation, or perhaps more appropriately as the Brahmagupta-Pell equation. A method for obtaining its solutions is described in Appendix O.

B.   Determine the number of valid ways of *sprinkling* labels from an allowed sequence onto the edges. This can be mapped to one of the simpler examples of NP-complete problems in the field of computational complexity—the number partitioning problem (NPP) [10,11].

The relevance of the NPP for black hole entropy arises as follows: The counting of states of a horizon for a non-rotating black hole boils down to determining the number of ways in which we can choose spin-labels k_i from a given sequence $\{k_1, ..., k_N\}$ (where the allowed sequences are determined by solving the Brahmagupta-Pell equation) to each of the $i = 1...N$ edges puncturing the horizon, such that $\sum k_i = 0$.

More generally the case where $\sum k_i = m$ ($m > 0$), corresponds to a horizon with angular momentum m. This is equivalent to the statement of the NPP, where given an arbitrary but fixed sequence of (positive) integers $A = \{a_i, ..., a_N\}$, one asks for the number of ways N_A in which we can partition A into two subsets A_+ and A_-, such that the difference of the sum of the elements of each subset is minimized, $\sum_{A_+} a_i - \sum_{A_-} a_i = m$. For the black hole entropy problem m is given by $\sum k_i^+ - \sum k_i^- = m$.

As shown in [8] this problem can be mapped to a non-interacting spin-system [12] as follows. Consider a chain of N spins each of which can be in an up $|\uparrow\rangle$ state or a down $|\downarrow\rangle$ state. If a_i belongs to A_+ (A_-) then we set the ith spin to up (down). Consequently the constraint $A_+ - A_- = m$ can be expressed as the condition that

$$m - \sum_{i=1}^{N} a_i S_i = 0 \tag{8.14}$$

where $S_i \in \{+1, -1\}$ are the possible eigenvalues of σ_z. The problem of partitioning A is then equivalent [12] to determining the ground state of the Hamiltonian

$$H = m - \sum_{i=1}^{N} a_j \sigma_z^j \qquad (8.15)$$

where σ_z^j is the Pauli spin operator for the jth spin. Any eigenstate of H with zero energy corresponds to a solution of the NPP for the set A.

8.1.3 Chern-Simons Approach

Another approach to the black hole entropy problem rests on the observation that the dynamics of punctures on the black hole horizon, in the framework of LQG, is described by a Chern-Simons theory. This relationship was first observed in the classic papers by Ashtekar et al. [13–15].

Building upon these findings, Kaul and Majumdar [16] were the first to show that the Bekenstein-Hawking expression for the entropy of a four-dimensional Schwarzschild black hole could be obtained from the dimensionality of the Hilbert space of an SU(2) Chern-Simons theory living on the horizon of that black hole. In addition to the leading term proportional to the area of the horizon, they were also able to obtain corrections proportional to the logarithm of the area in [17]. For a recent updated review of their findings see [18, 19].

More recently work by Engle and co-workers [20,21], reaches a similar conclusion by working in a manifestly SU(2) invariant formulation of the horizon degrees of freedom.

For a possible connection between the physics of the quantum Hall effect, as described by a Chern-Simons theory, and the question of black hole entropy see [22,23].

In this regard it is also worth mentioning that Chern-Simons theory provides an exact solution of the Hamiltonian constraint in terms of Ashtekar's self-dual variables. This solution is known as the Kodama state [24] and its properties have been extensively studied by Randono [25–27]. A brief introduction to the Kodama state is given in Appendix G.

8.1.4 Entropy from Entanglement

The relationship between entropy of entanglement and the entropy of black hole horizons was first suggested more than two decades ago in [28,29].

Consider a system, which could be a one-dimensional spin-chain (Fig. 8.2a) or a quantum field theory living on some spacetime (Fig. 8.2b). Divide the system into two parts A and B. Let ρ_{AB} be the density matrix representing the state of the system

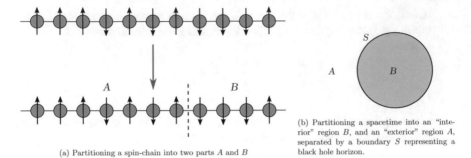

(a) Partitioning a spin-chain into two parts A and B

(b) Partitioning a spacetime into an "interior" region B, and an "exterior" region A, separated by a boundary S representing a black hole horizon.

Fig. 8.2 Entropy of entanglement is obtained by tracing over the degrees of freedom in either A or B

as a whole and ρ_A, ρ_B be the density matrices for systems A, B obtained by tracing over the degrees of freedom of A, B respectively, where

$$\rho_A = \text{Tr}_B[\rho_{AB}], \qquad \rho_B = \text{Tr}_A[\rho_{AB}]. \tag{8.16}$$

The von-Neumann entropy S_A of region A (or the entropy S_B of region B) can now be defined as

$$S_A = \text{Tr}_A [\rho_A \ln \rho_A], \qquad S_B = \text{Tr}_B [\rho_B \ln \rho_B]. \tag{8.17}$$

The two entropies are equal, $S_A = S_B$, so we only need to calculate one of them. In [28,29] a scalar field was employed as a "probe" field. After performing the trace over the interior region (B in Fig. 8.2b), the reduced density matrix ρ_A was found. Its von-Neumann entropy was found to be proportional to the area of the boundary surface S separating the interior and exterior regions.[5] This relationship between the entanglement entropy and the "area" of the boundary has turned out to be very general and is not limited to $3 + 1$ dimensional spacetime or to scalar fields. Some reviews of this phenomenon of "holographic" entanglement entropy are [30,31]. A perspective inspired by the anti-de Sitter/Conformal Field Theory (AdSCFT) correspondence can be found in [32].

   In 2010, a seminal paper by Mark Van Raamsdonk [33] argued that entanglement is the glue that holds spacetime together. A similar idea had earlier been suggested by Brian Swingle [34] who proposed that holographic spacetimes find a realization in the structure of networks formed from a technique, originally developed for studying many body strongly correlated systems, known as *entanglement renormalization* [35,36]. Since then a substantial amount of work has been done [37,38] towards providing concrete support for this proposal.

---

[5] Strictly speaking, a surface and its area are not the same thing, and the entropy is proportional to the area rather than equal to it. However we feel it is acceptable in this case to use the standard symbol for entropy, S, to indicate the boundary surface between the two regions, because of the conceptual connection between the surface and the entropy.

In 2006 Livine and Terno [39] first suggested that entanglement between different parts of the horizon might contribute to the Bekenstein-Hawking entropy of the black hole. In 2008 Donelly [40] and in 2012 Bianchi and Myers [41,42] proposed that Bekenstein-Hawking entropy could be understood as arising from the entropy of entanglement between the quantum geometric degrees of freedom on either side of the horizon. Similar ideas have been put forth by Dasgupta [43]. While these proposals have much in common with Raamsdonk & Swingle's ideas of geometry emerging from entanglement, the question of black hole entropy has yet to be addressed in the context of entanglement renormalization.

In conclusion, the study of black hole entropy in LQG is a very rich and active field and the results presented in this review, while very important and pioneering, should not been seen as the final word on this topic.

8.2   Loop Quantum Cosmology

One of the first avenues to follow when approaching old problems with new tools is to select the simplest possible scenarios for study, in the hope that the understanding gained in this arena would ultimately lead to a better understanding of more complex systems and processes. In classical GR this corresponds to studying the symmetry reduced solutions[6] of Einstein's equations, such as the Friedmann-LeMaitre-Robertson-Walker (FLRW) cosmologies and their anisotropic counterparts, and various other exact solutions such as deSitter, anti-deSitter, Schwarzschild, Kerr-Newman etc.[7] which correspond respectively to a "universe" (in this very restricted sense) with positive cosmological constant ($\Lambda > 0$), a universe with $\Lambda < 0$, a non-rotating black hole and a rotating black hole (both in asymptotically flat space-times[8]). In each of these cases the metric has a very small number of local degrees of freedom and hence provides only a "toy model". Of course, in the *real world*, the cosmos is a many-body system and reducing its study to a model such as the FLRW universe is a gross simplification. However, via such models, one can obtain a qualitative grasp of the behavior of the cosmos on the largest scales. This is the spirit in which the ideas of LQG are applied to the cosmos as a whole, leading to the field of Loop Quantum Cosmology (LQC).

The following discussion draws primarily from [46–48]. For a more in-depth introduction to the topic the reader is invited to consult [49,50].

---

[6] That is, the solutions of the EFEs possessing strong global symmetries which reduces the effective local degrees of freedom to a small number.

[7] We refer the reader to the extremely comprehensive and well-researched catalog of solutions to Einstein's field equations, in both metric and connection variables, presented in [44]. A somewhat older, but still valuable, catalog of exact solutions is given in [45].

[8] A metric with a radial dependence is considered asymptotically flat if it approaches (in a well-defined sense) a flat Minkowski metric as $r \to \infty$.

8.2.1 Isotropy and Homogeneity in the Metric Formulation

In the metric formulation the statements of isotropy and homogeneity of a spacetime are as follows:

Homogeneity A given spacetime geometry $\mathcal{M}_g = (\mathcal{M}, g_{\mu\nu})$, consisting of a manifold $\mathcal{M}$ and a metric defined on that manifold, $g_{\mu\nu}$, is said to be *spatially homogeneous* [48, Sect. 4.1.1] if there exists a symmetry group S acting on spatial slices Σ_t, such that for any two points $x, y \in \Sigma_t$, there exists an $s \in S$ such that $s(x) = y$. Under the action of s, the spatial metric h_{ab} on Σ_t, satisfies

$$h_{ab}(x) = s^\star h_{ab}(x) = h_{ab}(y) \tag{8.18}$$

where $s^\star$ is the pullback[9] of the spatial metric under the action of the isometry s.

Isotropy A given spacetime geometry $\mathcal{M}_g$ is said to be *isotropic* if at any point $x^\mu \in \mathcal{M}$, the metric satisfies

$$g_{\mu\nu}(x)u^\mu v^\nu = g_{\mu\nu}(x)(Ru)^\mu (Rv)^\nu \tag{8.19}$$

where u^μ, v^ν are arbitrary vectors in the tangent space $T_p(x)$ at that point and R is an arbitrary rotation acting on elements of $T_p(x)$.

Isotropy is a more restrictive condition than homogeneity, because isotropy necessarily implies homogeneity, however the reverse is not true.

8.2.2 FLRW Models

The simplest quantum cosmological model is that which corresponds to the Friedmann metric whose line-element is given by[10]

$$ds^2 = -N(t)^2 dt^2 + a(t)^2 \left(\frac{1}{1 - kr^2} dr^2 + r^2 d\Omega^2 \right) \tag{8.20}$$

where the only dynamical variable is the scale factor $a(t)$ which depends only on the time parameter, $r = \sqrt{x^2 + y^2 + z^2}$ is the radial dimension of the spatial slices, $d\Omega^2 = d\theta^2 + \sin^2\theta d\phi^2$ is the angular volume element and $k = -1, 0, +1$ determines whether our spatial slices are open ($k = -1$), flat ($k = 0$) or closed ($k = 1$). For this metric we can perform the $3 + 1$ decomposition into a foliation of spatial

---

[9] If r is a smooth map from one manifold to another $r : M \to N$, and f is a smooth function $f : N \to P$, the pullback of f by r, denoted $r^\star f$, is the map from M to P such that $(r^\star f)(x) = f(r(x))$ where x is in M. In the case given in the text s is a map from the manifold Σ_t to itself.

[10] The following discussion is taken from [47, Sect. 4].

manifolds Σ_t, and write down the action in terms of the various constraints. By comparing this metric with the general form given in Eq. (4.15), we see that $N(t)$ is the lapse function and the shift vanishes, $N^a = 0$. This implies that the diffeomorphism constraint $D_a \pi^{ab}$ must also vanish.

Inserting this metric into the the EFE (2.12) gives us the vacuum FLRW equations which describe the dynamics of homogenous, isotropic spacetimes

$$\left(\frac{\dot{a}}{a}\right)^2 + \frac{k}{a^2} = \frac{8\pi\mathcal{G}}{3a^2} H_{\text{matter}}(a) \tag{8.21}$$

where H_{matter} is the Hamiltonian for any matter fields that might be present. This equation gives us the Hamiltonian constraint for the FLRW metric. This can be seen by starting from the Lagrangian formulation where

$$S_{\text{EH}} = \frac{1}{16\pi\mathcal{G}} \int dt \, d^3x \, \sqrt{-g} R[g] . \tag{8.22}$$

The Ricci scalar $R[g]$ for the FLRW line-element (8.20) is

$$R = 6\left(\frac{\ddot{a}}{N^2 a} + \frac{\dot{a}^2}{N^2 a^2} + \frac{k}{a^2} - \frac{\dot{a}\dot{N}}{a N^3}\right) . \tag{8.23}$$

Substituting the above into the S_{EH} we obtain

$$S = \frac{V_0}{16\pi\mathcal{G}} \int dt \, N a(t)^2 R = \frac{3V_0}{8\pi\mathcal{G}} \int dt \, N \left(-\frac{a\dot{a}^2}{N^2} + ka\right) \tag{8.24}$$

where $V_0 = \int_{\Sigma} d^3x$ is the volume of a *fiducial cell* $\mathcal{V}$ in the spatial manifold. From this equation we can identify the momentum p_a conjugate to the (only) degree of freedom—the scale factor $a(t)$:

$$p_a = \frac{\partial L}{\partial \dot{a}} = -\frac{3V_0}{4\pi\mathcal{G}} \frac{a\dot{a}}{N} . \tag{8.25}$$

Since the action does not contain any terms depending on $\dot{N}$, we have $p_N = 0$, implying that the lapse function $N(t)$ is not a dynamical degree of freedom. We can now write down the Hamiltonian for the system in the usual manner, $H = \sum_i p_i \dot{q}_i - L = p_a \dot{a} - L$, which gives

$$H_{\text{grav}} = -N\left[\frac{2\pi\mathcal{G}}{3V_0} \frac{p_a^2}{a} + \frac{3V_0}{8\pi\mathcal{G}} ka\right] . \tag{8.26}$$

It is clear from the form of this expression that this Hamiltonian will become divergent as $a \to 0$. Changing from metric to connection variables will allow us to alleviate this problem.

8.2.3 Connection Variables

In the connection formulation the definition of isotropy and homogeneity is different from that in the metric picture, because here the relevant variables—the connection and tetrad—transform not under the action of diffeomorphisms, but under the action of gauge transformations g.[11]

Gauge Transformation Given a gauge group G (typically $SL(2, \mathbb{C})$ for four-dimensional Lorentzian gravity or $SU(2)$ for the three-dimensional spatial slices) and a manifold $\mathcal{M}$, a gauge transformation is a map $g : \mathcal{M} \to G$ from the manifold to the group, whose action on a given connection-tetrad (or triad) pair (A_a^i, e_i^a) on $\mathcal{M}$, is given by

$$(A', e') = (g^{-1}Ag + g\,dg, g^{-1}eg). \tag{8.27}$$

Homogeneity and Isotropy A given connection-tetrad/triad pair (A_a^i, e_i^a) on a manifold $\mathcal{M}$, is said to be spatially homogenous and isotropic [46, Sect. 7.1] if $\mathcal{M}$ is equipped with an isometry group S, and for every $s \in S$ there exists a gauge transformation $g : \mathcal{M} \to G$ such that

$$(s^\star A, s^\star e) = (g^{-1}Ag + g\,dg, g^{-1}eg). \tag{8.28}$$

Let us fix a *fiducial* flat metric $^0h_{ab}$ on $\mathcal{M}$ and the associated tetrad $^0e_i^a$ and co-triad $^0\omega_a^i$. Then every symmetric pair (A', E') on $\mathcal{M}$, can always be written in the form

$$A_a = \tilde{c}\,{}^0\omega_a^i \tau_i, \quad e^a = \tilde{p}\,\sqrt{\det(^0h)}\,{}^0e_i^a \tau^i \tag{8.29}$$

by choosing a suitable local gauge transformation (8.28). Here $\tau_i = -\frac{1}{2}\sigma_i$, with σ_i being the Pauli matrices and generators of the Lie algebra $\mathfrak{su}(2)$.

Thus the only non-trivial information in the pair (A', E') is contained in the two c-numbers $(\tilde{c}, \tilde{p})$, in terms of which the connection and triad can be written as

$$A_a^i = \tilde{c}\,\delta_a^i, \quad e_i^a = \tilde{p}\,\delta_i^a. \tag{8.30}$$

In variables adapted to the particular form of the metric (8.20), the connection $\tilde{c}$ and triad $|\tilde{p}|$ are expressed as

$$|\tilde{p}| = \frac{a^2}{4}, \quad \tilde{c} = \tilde{\Gamma} + \gamma\dot{a} = \frac{1}{2}(k + \gamma\dot{a}) \tag{8.31}$$

where γ is the Immirzi parameter. The Poisson bracket between these variables is

$$\{\tilde{c}, \tilde{p}\} = \frac{8\pi G\gamma}{3}V_0. \tag{8.32}$$

---

[11] Gauge transformations for the case of abelian groups were discussed in Sect. 3.1. Here we cover the more general non-abelian case which arises in four-dimensional gravity.

The factors of V_0 can be absorbed into the definition of the variables to give us

$$c = V_0^{1/3} \tilde{c} \qquad p = V_0^{2/3} \tilde{p} \tag{8.33}$$

whose Poisson bracket is

$$\{c, p\} = \frac{8\pi \mathcal{G} \gamma}{3}. \tag{8.34}$$

In terms of these the Hamiltonian constraint (8.26) becomes (for the flat $k = 0$ cosmology)

$$H = -\frac{3}{8\pi \mathcal{G} \gamma^2} c^2 \, \text{sgn}(p) \sqrt{|p|} + H_{\text{matter}} = 0 \tag{8.35}$$

where the factor of $\text{sgn}(p)$ corresponds to the orientation of the triad, a feature that is lost in the metric framework.

8.2.4 Holonomy Variables

To proceed to quantum cosmology one can start with either (8.26) or (8.35). Working with the former, using a, p_a as the generalized co-ordinate and momentum respectively, one would obtain the Wheeler-deWitt equation. However, the WdW equation does not resolve the short-range singularity obtained in the limit $a \to 0$, since the momentum operator is expressed as a derivation with respect to the co-ordinate, $\hat{p}_a \sim \partial/\partial a$, which is a continuous operator and hence does not encode the discreteness of background geometry.

In the LQC literature, the holonomy of an $\mathfrak{su}(2)$ connection is usually expressed in terms of the matrices $\tau_j = -\frac{1}{2}\sigma_j$ as

$$g_e(A) = \mathcal{P} \exp\left(\int_e A_a^i \tau_i n^a(x) dx \right) \tag{8.36}$$

where $n^a(x)$ is the unit tangent vector to the curve at the point x^μ. For the isotropic connection variable $\tilde{c}$ (8.31), the holonomy can be evaluated along any straight curve of length $l = V_0^{1/3}$, to give

$$g_e(A) = \cos(lc/2)\mathbb{1} + 2\sin(lc/2)(n^a \tau_i{}^0 e_a^i) \tag{8.37}$$

where $^0e_a^i$ is the (constant) *fiducial triad* associated with the (constant) metric on spatial hypersurfaces[12]. For example, if n^a lies along the z-direction, then

$$g_e(A) = \begin{bmatrix} \cos(lc/2) - \mathbf{i}\sin(lc/2) & 0 \\ 0 & \cos(lc/2) + \mathbf{i}\sin(lc/2) \end{bmatrix} = \begin{bmatrix} e^{-\mathbf{i}lc/2} & 0 \\ 0 & e^{\mathbf{i}lc/2} \end{bmatrix} \tag{8.38}$$

---

[12] Note that the peculiar factor of 2 appearing in the second term on the right-hand side is a consequence of working with the matrices $\tau_j = -\frac{1}{2}\sigma_j$, rather than with σ_j.

where c is the rescaled connection variable, and not the speed of light! The matrix elements of the holonomy operators, for an isotropic homogeneous spacetime, acting in the fundamental representation of SU(2) will therefore be of the form $\exp(i\mu_j c)$, where j labels the edge along which the holonomy is evaluated, and μ_j depends on the length of that edge. In terms of these matrix elements, all states in the connection representation can be written in the form [48, Sect. 6.2.1.2]

$$\psi(c) = \sum_j f_j e^{i\mu_j c} \tag{8.39}$$

where $f_j \in \mathbb{C}, \mu_j \in \mathbb{R}$. The inner product between two such states is given by

$$\langle \psi_1 | \psi_2 \rangle = \lim_{T \to \infty} \int_{-T}^{T} dc\, \psi_1^* \psi_2 . \tag{8.40}$$

8.2.5 Quantisation

While states of the form $e^{i\mu_j c}$ look very much like the familiar plane-waves e^{ikx} of classical and (non-loop) quantum mechanics, in contrast to plane-waves holonomy states are discontinuous in the "momenta" μ_j. This can be understood by recognising that for general graph states, i.e. those not restricted to correspond to homogeneous and/or isotropic geometries, states living on different graphs are orthogonal (6.12),

$$\langle \Theta_{\Gamma'} | \Psi_\Gamma \rangle = \delta_{\Gamma, \Gamma'} . \tag{8.41}$$

In the present situation, when two states ψ_1 and ψ_2 are given in terms of two different sets of "momenta" $\{\mu_j\}$ and $\{\mu_{j'}\}$, the two states can be said to living on different graphs, and therefore are orthogonal whenever the sets of "momenta" for both states are not identical,

$$\langle \psi_1 | \psi_2 \rangle = 0, \quad \text{if } \{\mu_j\} \neq \{\mu_{j'}\} .$$

For states corresponding to individual edges, $\psi_i = e^{\mu_i c}$, we have

$$\langle \psi_i | \psi_j \rangle = \delta_{\mu_i, \mu_j} = \begin{cases} 1 & \text{if } \mu_i = \mu_j \\ 0 & \text{if } \mu_i \neq \mu_j \end{cases} \tag{8.42}$$

regardless of how small the difference $\mu_i - \mu_j$ is. Thus the basis states (8.39) for a homogeneous, isotropic spacetime are defined on a real number line, equipped not with the usual continuous topology, but with a *discrete* topology! Because of this fact there does not exist any operator corresponding to the connection $\hat{c}$.[13] Therefore

---

[13] One might hope that such an operator could be obtained by taking the derivative of $\widehat{\exp i\mu c}$ with respect to μ. This turns out to not be the case. For further details we recommend the reader to consult [46, Sect. 7.2.1, 7.2.2] and [47, Sect. 5.2.1].

when quantizing expressions involving powers of c, the corresponding operators have to be constructed from exponentials of the connection.

As an example, consider the factor of c^2, occurring in the definition of the Hamiltonian constraint (8.35). When constructing the operator for this constraint, we have to write $\hat{c}^2$ in terms of exponentials. This can be done by noting that

$$c^2 = \frac{\sin^2(\mu c)}{\mu^2} + \mathcal{O}(\mu^4)$$

and $\sin(\mu c)$ can in turn be expressed in terms of exponentials,

$$\sin(\mu c) = \frac{e^{\mathbf{i}\mu c} - e^{-\mathbf{i}\mu c}}{2\mathbf{i}},$$

which finally allows us to approximate the operator expression for $\hat{c}^2$ as

$$\hat{c}^2 = -\frac{\left(\widehat{\exp(\mathbf{i}\mu c)} - \widehat{\exp(-\mathbf{i}\mu c)}\right)^2}{4\mu^2} + \mathcal{O}(\mu^4) \tag{8.43}$$

or, equivalently

$$\hat{c}^2 = \frac{\widehat{\sin^2(\mu c)}}{\mu^2} + \mathcal{O}(\mu^4). \tag{8.44}$$

8.2.6   Triad Eigenstates and Volume Quantization

From the Poisson bracket relations (8.34) between the rescaled connection and triad variables (c, p), one can see that the commutator between the corresponding quantum operators will be

$$[\hat{c}, \hat{p}] = \frac{8\pi\gamma\mathcal{G}}{3}, \tag{8.45}$$

implying that, in the connection representation (where states are functions of the connections as in (8.39)), the operator $\hat{p}$, becomes a derivative with respect to c,

$$\hat{p} = (-\mathbf{i}\hbar)\frac{8\pi\gamma\mathcal{G}}{3}\frac{\partial}{\partial c} = -\frac{8\pi\gamma l_{\mathrm{P}}^2}{3}\frac{\partial}{\partial c}, \tag{8.46}$$

where in the second step we have absorbed the factor of $\mathcal{G}\hbar$ into the definition of the Planck length $l_{\mathrm{P}}^2 = \mathcal{G}\hbar$ (in units where the speed of light $c_{\mathrm{light}} = 1$). Then the action of $\hat{p}$ on the basis states $\psi_\mu(c) = \exp(\mathbf{i}\mu c)$ is

$$\hat{p}\psi_\mu = \frac{8\pi\gamma l_{\mathrm{P}}^2}{3}\mu\psi_\mu, \tag{8.47}$$

implying that the states $\psi_\mu(c) \equiv |\mu\rangle$ are eigenstates of the triad operator. We can now understand the meaning of the parameter μ. By noting that the physical volume of a unit cell is given in terms of the triad[14] as $V = |p|^{3/2}$, we can write down the action of the corresponding volume *operator* on the triad eigenstate

$$\widehat{V}|\mu\rangle = |\hat{p}|^{3/2}|\mu\rangle = \left(\frac{8\pi\gamma l_P^2}{3}\right)^{3/2}|\mu|^{3/2}|\mu\rangle , \tag{8.48}$$

and $|\mu|^{3/2}$ thus corresponds to the volume of a fiducial cell $\mathcal{V}$ of the spacetime when the "universe" (in the very restricted sense of LQC) is in the state $|\mu\rangle$.

8.2.7   Regularized FLRW Hamiltonian

By inserting the expression (8.44) for the operator $\widehat{c^2}$ in the expression (8.35), we obtain the loop regularized expression for the Hamiltonian operator corresponding to an isotropic, homogeneous, flat ($k = 0$) FLRW universe,

$$\widehat{H}_{\text{loop}} = -\frac{3}{8\pi\mathcal{G}\gamma^2}\frac{\widehat{\sin^2(\mu c)}}{\mu^2}\,\text{sgn}(p)\sqrt{|\hat{p}|} + \widehat{H}_{\text{matter}} + \mathcal{O}(\mu^4) \simeq 0. \tag{8.49}$$

Since there is no explicit time-dependence in this expression, corresponding to the absence of a natural "clock" variable in general relativity, in order to understand how this loop spacetime evolves we must introduce an auxiliary clock variable, a role which is typically played by a massless scalar field in the matter sector.

The Hamiltonian for a massless scalar field in an isotropic background is given by

$$H_\phi(a, \phi, \pi_\phi) = \frac{1}{2}|p|^{-3/2}\pi_\phi^2 + |p|^{3/2}V(\phi) \tag{8.50}$$

where a is the scale factor, (ϕ, π_ϕ) are the generalized co-ordinate and momenta variables for the scalar field respectively, p is the isotropic triad and $V(\phi)$ is an optional potential for the scalar. The complete classical Hamiltonian constraint for a massless scalar (with no potential term) in an isotropic background is then given by

$$H = H_{\text{grav}} + H_\phi = -\frac{3}{8\pi\mathcal{G}\gamma^2}c^2\,\text{sgn}(p)\sqrt{|p|} + \frac{1}{2}\frac{\pi_\phi^2}{|p|^{3/2}} \simeq 0 \tag{8.51}$$

---

[14] In terms of the spatial metric, the volume of a unit cell is given by $V = \sqrt{\det h} = \sqrt{\epsilon_{abc}\epsilon^{ijk}e_i^a e_j^b e_k^c}$ (6.30). Since for an isotropic spacetime $e_i^a = p\,\delta_i^a$ (8.30), we find that $V = |p|^{3/2}$.

and likewise the *full* quantum Hamiltonian operator, for gravity plus matter, is given by

$$\widehat{H} = \widehat{H}_{\text{loop}} + \widehat{H}_{\text{matter}} = -\frac{3}{8\pi \mathcal{G}\gamma^2}\frac{\widehat{\sin^2(\mu c)}}{\mu^2}\,\text{sgn}(p)\sqrt{|\hat{p}|} + \frac{1}{2}\frac{\widehat{1}}{|p|^{3/2}}\hat{\pi}_\phi^2 \simeq 0\,.$$

(8.52)

Understanding how to obtain solutions of this equation will take us too far afield for an introductory review, so here we will only summarize the main implications of this quantization.

8.2.8 Singularity Resolution and Bouncing Cosmologies

Expressing the operator for the connection in the form (8.43), rather than in the familiar form $\hat{c} = \partial/\partial p$ from "normal" quantum mechanics, has an important implication for the resulting equations of motion. In quantum mechanics, the operator for the squared momentum becomes (in the position representation)

$$\hat{p}^2|\psi\rangle \sim \frac{\partial^2}{\partial x^2}|\psi\rangle$$

leading to the usual Schrödinger equation, which is a *differential equation*. However, in the "new" *loop quantum mechanics*, there is no operator corresponding to the connection $\hat{c}$, when working in the triad representation! Thus a $\hat{c}^2$ term has to be approximated by the form given in (8.43)

$$\hat{c}^2 \approx -\frac{\left(\widehat{\exp(i\delta c)} - \widehat{\exp(-i\delta c)}\right)^2}{4\delta^2} + \mathcal{O}(\delta^4)$$

$$= -\frac{1}{4\delta^2}\left(\widehat{\exp(i\delta c)}^2 + \widehat{\exp(-i\delta c)}^2 - 2\right) + \mathcal{O}(\delta^4)\,.$$

To understand the action of this operator on a triad eigenstate $|\mu\rangle$, we need the action of the operator $\widehat{\exp(i\delta c)}$ on $|\mu\rangle$. This can be easily seen to be

$$\widehat{\exp(i\delta c)}|\mu\rangle = |\mu + \delta\rangle$$

(8.53)

because $|\mu\rangle$ is nothing more than $e^{i\mu c}$!

Now we can easily determine the action of $\hat{c}^2$ on $|\mu\rangle$, which is

$$\hat{c}^2|\mu\rangle \approx -\frac{1}{4\delta^2}\left(|\mu + 2\delta\rangle + |\mu - 2\delta\rangle - 2|\mu\rangle\right) + \mathcal{O}(\delta^4)\,.$$

(8.54)

One can see that instead of a differential equation, in the LQC approach, we will obtain *difference* equations. One might argue that if the limit $\delta \to 0$ is taken in the above expression, the left-hand side will reduce to the usual expression for the second derivative

$$\lim_{\delta \to 0} \frac{f(x+2\delta) + f(x-2\delta) - 2f(x)}{\delta^2} = \frac{d^2 f(x)}{dx^2}.$$

However, as (8.48) shows, $(8\pi\gamma/3)l_P^3 |\mu|^{3/2}$ corresponds to the volume of a fiducial "cell". As shown in Sect. 6.4, the action of the volume operator $\widehat{V}$ in full LQG - i.e. before any symmetry reduction has been performed—on a spin network state Ψ_Γ, living on the graph Γ, is of the form

$$\widehat{V}\Psi_\Gamma \sim \sum_{v \in S \cap \Gamma} \sqrt{\epsilon_{abc}\epsilon^{ijk}n^a n^b n^c \widehat{J}_i \widehat{J}_j \widehat{J}_k}$$

modulo some constants and choice of sign factors. Here S is the region of the manifold whose volume we wish to obtain, $v \in S \cap \Gamma$ is set of vertices v of Γ which lie within S, and n^a, n^b, n^c are tangent vectors to the edges of Γ which meet at v.

Now, we know that the operator $\widehat{J}$ is bounded from below, i.e. has a minimum eigenvalue of $1/2$. Thus the volume operator $\widehat{V}$ *must* also necessarily be bounded from below, with the smallest possible volume eigenvalue in LQG being of the order $\gamma^{3/2}l_P^3$.

Thus we arrive at the conclusion that because of the quantization of geometric operators in full LQG it is *not* permissible to take the limit $\delta \to 0$ in expressions such as (8.54). This fact lies at the core of the observation that LQC cures the singularities in cosmological evolution that are encountered in the limit that the scale factor $a \to 0$ when we solve the classical FRLW equations or their quantum counterparts, the Wheeler-deWitt equations.

Furthermore, a similar line of reasoning shows that the factor of $\widehat{|p|^{-3/2}}$, multiplying the scalar field momentum in the LQC Hamiltonian (8.52), remains bounded *from above* throughout the evolution of the universe. In classical general relativity, where $p \sim a^2$, as the initial Big Bang singularity is approached, the scale factor diverges $a \to 0$, leading to infinite energy densities for the scalar (and any other matter) field. In LQC, the fact that $\widehat{|p|^{-3/2}}$ has an upper bound of order $\sim l_P^{-3}$, ensures that such divergences do not occur. The consequence is that cosmological evolution remains regular and non-singular as one approaches the Big Bang and, in fact, one can evolve *past* the moment of creation into what can be interpreted as a collapsing phase of a universe which existed *before* our own!

Though the interpretation of the various branches of the cosmological evolution in LQC as bouncing universes might be a matter of some debate, it is clear that due to the tight consistency constraints on this geometric approach to quantum gravity we can rest assured that singularities such as the one encountered at the moment of the Big Bang or the one which is the end result of uncontrolled gravitational collapse of matter, resulting in formation of a black hole, are artifacts of a description of geometry which implicitly relied on the assumption of an infinitely smooth and continuous spacetime at all scales.

References

1. S. Hawking, Particle creation by black holes. Commun. Math. Phys. **43**(3), 199–220 (1975). https://doi.org/10.1007/BF02345020
2. L. Susskind, The world as a hologram. J. Math. Phys. **36**(11), 6377–6396 (1994). issn: 00222488. https://doi.org/10.1063/1.531249. arXiv:hep-th/9409089
3. G. 't Hooft, *Dimensional Reduction in Quantum Gravity* (1993). https://doi.org/10.48550/arXiv.gr-qc/9310026. arXiv:gr-qc/9310026. http://arXiv.org/abs/gr-qc/9310026
4. E.T. Jaynes, Information Theory and Statistical Mechanics. Phys. Rev. Online Archive (Prola) **106**(4), 620–630 (1957). https://doi.org/10.1103/PhysRev.106.620
5. C. E. Shannon, A mathematical theory of communication. Bell Syst. Techn. J. **27**, 379–423 & 623–656 (1948). https://doi.org/10.1002/j.1538-7305.1948.tb01338.x. https://people.math.harvard.edu/~ctm/home/text/others/shannon/entropy/entropy.pdf
6. R.K. Pathria, P.D. Beale, *Statistical Mechanics*, 3rd edn. (Academic Press, Mar. 2011). isbn: 0123821886. https://doi.org/10.1016/C2009-0-62310-2. https://www.sciencedirect.com/book/9780123821881/statisticalmechanics
7. C. Rovelli, Black hole entropy from loop quantum gravity. Phys. Rev. Lett. **77**, 3288–3291 (1996). https://doi.org/10.1103/PhysRevLett.77.3288. (arXiv:gr-qc/9603063)
8. I. Agullo, et al., Black hole state counting in loop quantum gravity: a number theoretical approach. Phys. Rev. Lett. **100**(21), 211301 (2008). issn: 0031-9007. https://doi.org/10.1103/PhysRevLett.100.211301. arXiv:0802.4077
9. I. Agullo et al., Detailed black hole state counting in loop quantum gravity. Phys. Rev. D **82**, 084029 (2010). https://doi.org/10.1103/PhysRevD.82.084029. (arXiv:1101.3660)
10. S. Mertens, Phase transition in the number partitioning problem. Phys. Rev. Lett. **81**(20), 4281–4284 (1998). issn: 0031-9007. https://doi.org/10.1103/physrevlett.81.4281. arXiv:cond-mat/9807077. http://dx.doi.org/10.1103/physrevlett.81.4281
11. S. Mertens, A physicist's approach to number partitioning. Theor. Comp. Sci. **265**, 79–108 (2000). https://doi.org/10.48550/arXiv.cond-mat/0009230. arXiv:cond-mat/0009230
12. H. De Raedt et al., Number partitioning on a quantum computer (2001). https://doi.org/10.1016/S0375-9601(01)00680-6. arXiv:quant-ph/0010018
13. A. Ashtekar et al., Quantum geometry and black hole entropy. Phys. Rev. Lett. **80**, 904–907 (1998). https://doi.org/10.1103/PhysRevLett.80.904. (arXiv:gr-qc/9710007)
14. A. Ashtekar, J. Baez, K. Krasnov, Quantum geometry of isolated horizons and black hole entropy. Adv. Theor. Math. Phys. **4**, 1–94 (2000). https://doi.org/10.48550/arXiv.gr-qc/0005126. arXiv:gr-qc/0005126
15. A. Ashtekar, A. Corichi, K. Krasnov, Isolated horizons: the classical phase space. Adv. Theor. Math. Phys. **3**, 419–478 (1999). https://doi.org/10.48550/arXiv.gr-qc/9905089. arXiv:gr-qc/9905089
16. R.K. Kaul, P. Majumdar, Quantum black hole entropy. Phys. Lett. B **439**, 267–270 (1998). https://doi.org/10.1016/S0370-2693(98)01030-2. (arXiv:gr-qc/9801080)
17. R.K. Kaul, P. Majumdar, Logarithmic correction to the Bekenstein-Hawking entropy. Phys. Rev. Lett. **84**, 5255–5257 (2000). https://doi.org/10.1103/PhysRevLett.84.5255. (arXiv:gr-qc/0002040)
18. R.K. Kaul, P. Majumdar, Schwarzschild horizon dynamics and SU(2) Chern-Simons theory. Phys. Rev. D **83**(2) (2010). issn: 1550-7998. https://doi.org/10.1103/PhysRevD.83.024038. arXiv:1004.5487. http://dx.doi.org/10.1103/PhysRevD.83.024038
19. R.K. Kaul, Entropy of quantum black holes. SIGMA (2012). issn: 18150659. https://doi.org/10.3842/SIGMA.2012.005. arXiv:1201.6102

20. J. Engle, K. Noui, A. Perez, Black hole entropy and SU(2) Chern-Simons theory. Phys. Rev. Lett. **105**, 031302 (2010). https://doi.org/10.1103/PhysRevLett.105.031302. (arXiv:0905.3168)
21. J. Engle et al., Black hole entropy from an SU(2)-invariant formulation of Type I isolated horizons. Phys. Rev. D **82**, 044050 (2010). https://doi.org/10.1103/PhysRevD.82.044050. (arXiv:1006.0634)
22. D. Vaid, Quantum hall effect and black hole entropy in loop quantum gravity (2012). https://doi.org/10.48550/arXiv.1208.3335. arXiv:1208.3335
23. A.G.A. Pithis, H.-C.R. Euler, Anyonic statistics and large horizon diffeomorphisms for Loop Quantum Gravity Black Holes (Feb. 2015). https://doi.org/10.1103/PhysRevD.91.064053. arXiv:1402.2274
24. H. Kodama, Holomorphic wave function of the Universe. Phys. Rev. D **42**, 2548–2565 (1990). https://doi.org/10.1103/PhysRevD.42.2548
25. A. Randono, Generalizing the Kodama state I: construction (2006). https://doi.org/10.48550/arXiv.gr-qc/0611073. arXiv:gr-qc/0611073
26. A. Randono, Generalizing the Kodama state II: properties and physical interpretation (2006). https://doi.org/10.48550/arXiv.gr-qc/0611074. arXiv:gr-qc/0611074
27. A. Randono, In search of quantum de sitter space: generalizing the Kodama state. Ph.D. thesis. University of Texas at Austin, 2007. https://doi.org/10.48550/arXiv.0709.2905
28. L. Bombelli, et al., Quantum source of entropy for black holes. Phys. Rev. D **34**(2), 373–383 (1986). issn: 0556-2821. https://doi.org/10.1103/physrevd.34.373
29. M. Srednicki, Entropy and area. Phys. Rev. Lett. **71**(5), 666–669 (1993). https://doi.org/10.1103/PhysRevLett.71.666. (arXiv:hep-th/9303048)
30. S.D.S. Shankaranarayanan, S. Sur, Black entropy from entanglement: a review. Horiz. World Phys. **268** (2009). (Ed. by M. Everett, L. Pedroza). https://doi.org/10.48550/arXiv.0806.0402. arXiv:0806.0402
31. S.N. Solodukhin, Entanglement entropy of black holes. Living Rev. Relativ **14**, 8 (2011). https://doi.org/10.12942/lrr-2011-8. arXiv:1104.3712
32. T. Nishioka, S. Ryu, T. Takayanagi, Holographic entanglement entropy: an overview. J Phys A Math Theor **42**(50) (2009), 504008+. issn: 1751-8113. https://doi.org/10.1088/1751-8113/42/50/504008. arXiv:0905.0932
33. M. Van Raamsdonk, Building up spacetime with quantum entanglement. Gen. Rel. Grav. **42**, 2323–2329 (2010). https://doi.org/10.1007/s10714-010-1034-0. (arXiv:1005.3035)
34. B. Swingle, Entanglement renormalization and holography. Phys. Rev. D **86**, 065007 (2012). https://doi.org/10.1103/PhysRevD.86.065007. (arXiv:0905.1317)
35. G. Vidal, Entanglement renormalization. Phys. Rev. Lett. **99**, 220405 (2007). https://doi.org/10.1103/PhysRevLett.99.220405. (arXiv:cond-mat/0512165)
36. G. Vidal, Entanglement renormalization: an introduction, in *Understanding Quantum Phase Transitions*, ed. by L.D. Carr (2010). https://doi.org/10.48550/arXiv.0912.1651. arXiv:0912.1651
37. B. Swingle, Constructing holographic spacetimes using entanglement renormalization (2012). https://doi.org/10.48550/arXiv.1209.3304. arXiv:1209.3304
38. B. Swingle, M. Van Raamsdonk, Universality of gravity from entanglement (2014). https://doi.org/10.48550/arXiv.1405.2933. arXiv:1405.2933
39. E.R. Livine, D.R. Terno, Quantum black holes: entropy and entanglement on the horizon. Nucl. Phys. B **741**(1–2), 131–161 (2006). issn: 05503213. https://doi.org/10.1016/j.nuclphysb.2006.02.012. arXiv:gr-qc/0508085
40. W. Donnelly, Entanglement entropy in loop quantum gravity. Phys. Rev. D 77(10) (2008). issn: 1550-7998. https://doi.org/10.1103/physrevd.77.104006. arXiv:0802.0880
41. E. Bianchi, R.C. Myers, On the architecture of spacetime geometry (2012). https://doi.org/10.48550/arXiv.1212.5183. arXiv:1212.5183
42. E. Bianchi, Black hole entropy from graviton entanglement (Jan. 2013). https://doi.org/10.48550/arXiv.1211.0522. arXiv:1211.0522
43. A. Dasgupta, Semiclassical loop quantum gravity and black hole thermodynamics. SIGMA 9 (2013). issn: 18150659. https://doi.org/10.3842/sigma.2013.013. arXiv:1203.5119

44. T. Mueller, F. Grave, Catalogue of spacetimes (2009). https://doi.org/10.48550/arXiv.0904. 4184. arXiv:0904.4184

45. H. Stephani, et al., Exact solutions of Einstein's field equations. (Cambridge Monographs on Mathematical Physics), 2nd edn. (Cambridge University Press, May 2003). isbn: 9780521461368. https://doi.org/10.1017/CBO9780511535185

46. A. Ashtekar, J. Lewandowski, Background independent quantum gravity: a status report. Class. Quant. Grav. **21**(15), R53–R152 (2004). https://doi.org/10.1088/0264-9381/21/15/R01. (arXiv:gr-qc/0404018)

47. M. Bojowald, Loop quantum cosmology. Living Rev. Relativ. **8**, 11. https://doi.org/10.12942/lrr-2005-11. arXiv:gr-qc/0601085

48. M. Bojowald, *Canonical Gravity and Applications: Cosmology, Black Holes, Quantum Gravity* (Cambridge University Press, 2010). https://doi.org/10.1017/CBO9780511921759

49. A. Ashtekar, P. Singh, Loop quantum cosmology: a status report. Class. Quant. Grav. **28**(21), 213001 (2011). https://doi.org/10.1088/0264-9381/28/21/213001. (arXiv:1108.0893)

50. K. Banerjee, G. Calcagni, M. Martín-Benito, Introduction to loop quantum cosmology. SIGMA **8**, 016 (2012). https://doi.org/10.3842/SIGMA.2012.016. (arXiv:1109.6801)

Discussion

9

Any fair and balanced review of LQG should also mention at least a few of the many objections its critics have presented. A list of a few of the more important points of weakness in the framework and brief responses to them follows:

1. *LQG admits a volume extensive entropy and therefore does not respect the Holographic principle*: This criticism hinges upon the description of states of quantum gravity as spin networks which are essentially spin-systems on arbitrary graphs. However, spin networks only constitute the *kinematical* Hilbert space of LQG. They are solutions of the spatial diffeomorphism and the Gauss constraints but *not* of the Hamiltonian constraint which generates time-evolution. This criticism is therefore due to a (perhaps understandable) failure to grasp the difference between the kinematical and the dynamical phase space of LQG.

    In order to solve the Hamiltonian constraint we are forced to enlarge the set of states to include *spin foams* which are histories of spin networks. In a nutshell then, as we mentioned in Sect. 6.6, the kinematical states of LQG are the spin networks, while the dynamical states are the spin foams. The amplitudes associated with a given spin foam are determined completely by the specification of its boundary state. Physical observables do not depend on the possible internal configurations of a spin foam but only on its boundary state. In this sense LQG satisfies a stronger and cleaner version of holography than string theory, where this picture emerges from considerations involving graviton scattering from certain extremal black hole solutions.

    In [1] it is shown that in the context of loop quantum cosmology of a radiation-filled flat FLRW model, Bousso's covariant entropy bound [2] is respected. As one approaches the moment of the Big Bang, and quantum gravitational effects become large the bound is violated, however, far from the Big Bang, when geometry has become semiclassical the bound comes into force.

    As yet, there is no general proof of whether or not LQG respects Bousso's bound. However, one might argue that the structure of LQG is amenable to the spirit of

© Springer Nature Switzerland AG 2024
S. Bilson-Thompson, *Loop Quantum Gravity for the Bewildered*,
https://doi.org/10.1007/978-3-031-43452-5_9

Bousso's bound. The latter suggests that there is a fundamental limit to the number of degrees of freedom in any given region of spacetime. Such a fundamental limit is already present in LQG in the form of the quantized area and volume operators which tell us that any region of spacetime must contain a finite number of geometric degrees of freedom.

2. *LQG violates the principle of local Lorentz invariance/picks out a preferred frame of reference*: Lorentz invariance is obeyed in LQG but obviously not in the exact manner as for a continuum geometry. As has been shown by Rovelli and Speziale [3] the kinematical phase space of LQG can be cast into a manifestly Lorentz covariant form. A spin network/spin foam state transforms in a well-defined way under boosts and rotations. Similarly in quantum mechanics one finds that a quantum rotor transforms under discrete representations of the rotation group SO(3).

3. *LQG does not have stable semiclassical geometries as solutions—geometry "crumbles"*: CDT simulations e.g. [4] show how a stable geometry emerges. As mentioned in Sect. 4.1, this involves calculating a sum over histories for the geometry of spacetime, between some initial and final state. The stability of the spacetimes studied in such simulations appears to be dependent on causality—that is, spacetime geometries develop unphysical structures in the Euclidean case, which are controlled when there is a well-defined past and future, as is the case in LQG. The question of exactly how similar CDT and LQG are to each other is a matter of continuing investigation.

4. *LQG does not contain fermionic and bosonic excitations that could be identified with members of the Standard Model*: The area and volume operators do not describe the entirety of the structures that can occur within spin networks. LQG or a suitably modified version which allows braiding between various edges will exhibit invariant topological structures. Recent work [5–10] has been able to identify some such structures with SM particles. In addition, in any spin-system—such as LQG—there are effective (emergent) low-energy degrees of freedom which satisfy the equations of motion for Dirac and gauge fields. Xiao-Gang Wen and Michael Levin [11, 12] have investigated so-called "string-nets" and find that the appropriate physical framework is the so-called "tensor category" or "tensor network" theory [13–15]. In fact string-nets are very similar to spin networks so Wen and Levin's work—showing that gauge bosons and fermions are quasiparticles of string-net condensates—should carry over into LQG without much modification.

5. *LQG does not exhibit dualities in the manner String Theory does*: Any spin system exhibits dualities. A graph based model like LQG even more so. One example of a duality is to consider the dual of a spin network which is a so-called 2-skeleton or simplicial cell-complex. Another is the star-triangle transformation, which can be applied to spin networks which have certain symmetries, and which leads to a duality between the low and high temperature versions of a theory on a hexagonal and triangular lattice respectively [16].

6. *LQG doesn't admit supersymmetry, wants to avoid extra dimensions, strings, extended objects, etc*: Extra dimensions and supersymmetry are precisely that—

"extra". Occam's razor dictates that a successful physical theory should be founded on the *minimum* number of ingredients. It is worth noting that at the time of writing of this book, results from the Large Hadron Collider appear to have ruled out many supersymmetric extensions of the standard model. By avoiding the inclusion of extra dimensions and supersymmetry, LQG represents a perfectly valid attempt to create a theory that is consistent with observations.

7. *LQG has a proliferation of models and lacks robustness*: Again a lack of extra baggage implies the opposite. LQG is a tightly constrained framework. There are various uniqueness theorems which underlie its foundations and were rigorously proven in the 1990s by Ashtekar, Lewandowski and others. There are questions about the role of the Immirzi parameter and the ambiguity it introduces however these are part and parcel of the broader question of the emergence of semi-classicality from LQG (see Simone Mercuri's papers [17, 18] in this regard).

8. *LQG does not contain any well-defined observables and does not allow us to calculate graviton scattering amplitudes*: Several calculations of two-point correlation functions in spin foams exist in the literature [19]. These demonstrate the emergence of an inverse-square law.

As well as discussing criticisms of LQG, it is also fair to consider what role this theory may have in the future. We would not have written a review of the formulation of LQG if we did not consider it an important and interesting theory—one which we feel is probably a good representation of the nature of spacetime. However it is wise to remember that most physical theories are ultimately found to be flawed or inadequate representations of reality, and it would be unrealistic to think that the same might not be true of LQG. Questions linger about the nature of time and the interpretation of the Hamiltonian constraint, among other things. What is the value then, in studying LQG? Perhaps LQG will eventually be shown to be untenable, or perhaps it will be entirely vindicated. As authors of this book, we feel that the truth will probably lie somewhere in the middle, and that however much of our current theories of LQG survive over the next few decades, this research program does provide strong indications about what some future (and, we hope, experimentally validated) theory of "quantum gravity" will look like.

References

1. A. Ashtekar, E. Wilson-Ewing, Covariant entropy bound and loop quantum cosmology. Phys. Rev. D **78**(6), 064047 (2008). https://doi.org/10.1103/PhysRevD.78.064047. arXiv:0805.3511
2. R. Bousso, A covariant entropy conjecture. JHEP **1999**(07), 004 (1999). https://doi.org/10.1088/1126-6708/1999/07/004. arXiv:hep-th/9905177
3. C. Rovelli, S. Speziale, Lorentz covariance of loop quantum gravity. Phys. Rev. D **83**, 104029 (2011). https://doi.org/10.1103/PhysRevD.83.104029. arXiv:1012.1739
4. R. Loll, J. Ambjorn, J. Jurkiewicz, The universe from scratch. Contemp. Phys. **46**, 103–117 (2006). https://doi.org/10.1080/00107510600603344. arXiv:hep-th/0509010v3
5. S. Bilson-Thompson, *A Topological Model of Composite Preons* (2005). https://doi.org/10.48550/arXiv.hep-ph/0503213. arXiv:hep-ph/0503213

6. S.O. Bilson-Thompson, F. Markopoulou, L. Smolin, Quantum gravity and the standard model. Class. Quantum Grav. **24**, 3975–3994 (2007). https://doi.org/10.1088/0264-9381/24/16/002. arXiv:hep-th/0603022

7. S. Bilson-Thompson et al., *Particle Identifications from Symmetries of Braided Ribbon Network Invariants* (2008). https://doi.org/10.48550/arXiv.0804.0037. arXiv:0804.0037

8. S. Bilson-Thompson, J. Hackett, L.H. Kauffman, Particle topology, braids, and braided belts. J. Math. Phys. **50**, 113505 (2009). https://doi.org/10.1063/1.3237148. arXiv:0903.1376

9. S. Bilson-Thompson et al., Emergent braided matter of quantum geometry. SIGMA **8**, 014 (2012). https://doi.org/10.3842/SIGMA.2012.014. arXiv:1109.0080

10. D. Vaid, Embedding the Bilson-Thompson model in an LQG-like framework (2010). https://doi.org/10.48550/arXiv.1002.1462. arXiv:1002.1462

11. M.A. Levin, X.-G. Wen, String-net condensation: a physical mechanism for topological phases. Phys. Rev. B **71**, 045110 (2005). https://doi.org/10.1103/PhysRevB.71.045110. arXiv:cond-mat/0404617

12. M. Levin, X.-G. Wen, Detecting topological order in a ground state wave function. Phys. Rev. Lett. **96**, 110405 (2006). https://doi.org/10.1103/PhysRevLett.96.110405. arXiv:cond-mat/0510613

13. J.D. Biamonte, S.R. Clark, D. Jaksch, Categorical tensor network states. AIP Adv. **1**, 042172 (2011). https://doi.org/10.1063/1.3672009. arXiv:1012.0531

14. G. Evenbly, G. Vidal, Tensor network states and geometry. J. Stat. Phys. **145**, 891–918 (2011). https://doi.org/10.1007/s10955-011-0237-4. arXiv:1106.1082

15. J. Haegeman et al., Entanglement renormalization for quantum fields. Phys. Rev. Lett. 110, 100402 (2013). https://doi.org/10.1103/PhysRevLett.110.100402. arXiv:1102.5524

16. R.J. Baxter, *Exactly Solved Models in Statistical Mechanics* (Dover Publications, 2008). isbn: 0486462714. https://physics.anu.edu.au/research/ftp/mpg/baxter_book.php

17. S. Mercuri, From the Einstein-Cartan to the Ashtekar-Barbero canonical constraints, passing through the Nieh-Yan functional. Phys. Rev. D **77**(2) (2008). issn: 1550-7998. https://doi.org/10.1103/physrevd.77.024036. arXiv:0708.0037

18. S. Mercuri, Peccei-Quinn mechanism in gravity and the nature of the Barbero-Immirzi parameter. Phys. Rev. Lett. **103**(8) (2009). issn: 0031-9007. https://doi.org/10.1103/PhysRevLett.103.081302. arXiv:arXiv:0902.2764

19. C. Rovelli, Graviton propagator from background-independent quantum gravity. Phys. Rev. Lett. **97**, 151301 (2006). https://doi.org/10.1103/PhysRevLett.97.151301. arXiv:gr-qc/0508124

Groups, Representations, and Algebras

A

A.1 Lie Groups and Algebras

Symmetries in physics (that is to say, transformations which leave the underlying physics unchanged) and conserved quantities are intimately related by Noether's theorem. Symmetries play a pivotal role in establishing which physical laws are allowed. For instance we saw in Sect. 3.1 how the requirement that the laws of physics remain unchanged under a local gauge transformation necessitated the existence of a covariant derivative and a gauge field.

Symmetries have several properties which are taken as the defining features of a group. Consider a set G containing elements g_1, g_2, ... Furthermore, consider an operation, denoted here by the symbol $\circ$, which maps exactly two elements of G to some quantity. We will call such a mapping a binary operation. Then the set and binary operation taken together form a group if for all $g \in G$ the following properties are satisfied:

1. $g_j \circ g_k \in G$, $\forall g_j, g_k \in G$ (closure)
2. $\exists e \in G$ such that $g \circ e = e \circ g = g$ (identity)
3. $\forall g \in G \, \exists g^{-1} \in G$ such that $g^{-1} \circ g = g \circ g^{-1} = e$ (inverse)
4. $g_j \circ (g_k \circ g_l) = (g_j \circ g_k) \circ g_l \; \forall g_j, g_k, g_l \in G$ (associativity)

The groups we will be concerned with have elements that vary continuously as functions of some parameter or parameters. These are known as Lie groups. For example, rotations through some angle θ, which can be written as $g(\theta) = e^{i\theta}$, and which transform one point on a unit circle to another. In what follows we will call this the circle group. It is worth emphasising that the group elements in this case are the transformations of the points on a circle, and not the points themselves.

Rotations in a circle can also be enacted by operating on a vector by a 2×2 matrix. Such matrices have two important properties. Their transpose is also their inverse (we refer to such matrices as "orthogonal") and their determinant is equal to +1 (we

© Springer Nature Switzerland AG 2024
S. Bilson-Thompson, *Loop Quantum Gravity for the Bewildered*,
https://doi.org/10.1007/978-3-031-43452-5

refer to such matrices as "special"). Hence these matrices are called SO(2) matrices (short for "special orthogonal two-by-two"). Such matrices are again continuous functions of the single parameter θ, and are hence isomorphic to the circle group. The circle group is actually a 1×1 example of a "unitary" group (one whose transpose complex conjugate is its inverse). Hence the circle group is also called U(1). Note that the special unitary group of 1×1 matrices SU(1) consists of only a single element, the identity, and is referred to as the trivial group. While the special groups correspond to rotations, the groups of matrices with determinant $= -1$ correspond to reflections. General $N \times N$ matrices with arbitrary non-zero determinants (i.e. they are invertible) form the "general linear" group GL(N). Where required, the field of numbers used as matrix elements can be specified explicitly. For instance the group whose elements are $N \times N$ matrices with complex entries and arbitrary determinants is GL($N,\mathbb{C}$). The subgroup with determinant $= +1$ of this general linear group is SL($N,\mathbb{C}$). The unitary subgroup of this special linear group is SU(N), or SU($N,\mathbb{C}$) if one is being explicit. Restricting to the field of real numbers we obtain the special orthogonal group SO(N). The orthogonal groups represent rotations in a space with a Riemannian metric (i.e. one which always ascribes a separation greater than zero to distinct points). In Minkowski space, with metric $\eta_{\mu\nu} = \text{diag}(-1, +1, +1, +1)$, the group of Lorentz transformations is SO(3,1), denoting that in n dimensions (in this case $n = 4$), only $n - 1$ of the terms in the metric are positive.

Since SO(2) and U(1) both embody rotations in a circle, it is clear that there can be different *representations* of a group. In fact, we could represent rotations in the complex plane by an infinite number of different functions, $\exp\{ik\theta\}$, with each function being labelled by different integer[1] values of the parameter k. Specifically, a representation of a group is a mapping from group elements g to functions $D(g)$, which are usually taken to be matrices, such that $D(g_a)D(g_b) = D(g_a \circ g_b)$ for all a, b. We can easily see that, for instance, rotations in the plane can be represented by the 1×1 "matrices" $D_1(\theta) = e^{i\theta}$, and the 2×2 matrices

$$D_2(\theta) = \begin{pmatrix} \cos\theta & -\sin\theta \\ \sin\theta & \cos\theta \end{pmatrix}.$$

Some representations contain a level of redundancy, allowing them to be viewed as composed of simpler representations. Such redundant representations are referred to as being *reducible*. A representation which is not reducible is called an irreducible representation or *irrep*.

The tangent space to a Lie group (e.g. the group of rotations $g(\theta) = e^{i\theta}$) at the identity, is the corresponding Lie algebra. Rotations in three dimensions can be represented by the group SO(3), or the group SU(2). To be specific, consider a general element $R \in$ SU(2) which can be written in terms of three parameters $\vec{\beta} := (\beta_1, \beta_2, \beta_3)$ as

---

[1] So that $\exp\{ik2\pi\}$ is a representation of the identity.

$$R(\vec{\beta}) = \begin{pmatrix} \beta_4 + \mathbf{i}\beta_1 & \beta_2 + \mathbf{i}\beta_3 \\ -\beta_2 + \mathbf{i}\beta_3 & \beta_4 - \mathbf{i}\beta_1 \end{pmatrix} \quad \text{where } \beta_4 = +\sqrt{1 - \beta_1^2 - \beta_2^2 - \beta_3^2} \quad \text{(A.1)}$$

since $\det R = \beta_1^2 + \beta_2^2 + \beta_3^2 + \beta_4^2 = +1$. The identity corresponds to $\beta_4 = 1$, $\beta_1 = \beta_2 = \beta_3 = 0$, so the tangent space to SU(2) is spanned by

$$J_a = -\mathbf{i} \left. \frac{\partial R(\vec{\beta})}{\partial \beta_a} \right|_{\vec{\beta}=0, \beta_4=1} \quad \text{(A.2)}$$

$$\therefore J_1 = \begin{pmatrix} 1 & 0 \\ 0 & -1 \end{pmatrix}, \quad J_2 = \begin{pmatrix} 0 & -\mathbf{i} \\ \mathbf{i} & 0 \end{pmatrix}, \quad J_3 = \begin{pmatrix} 0 & 1 \\ 1 & 0 \end{pmatrix}. \quad \text{(A.3)}$$

Lie algebras are referred to with the same notation, but lower case fraktur lettering, as the corresponding group. Hence the Lie algebra of SU(2) is $\mathfrak{su}(2)$. Since this group corresponds to rotations in three dimensions, $\mathfrak{su}(2)$ has three generators, which we recognise from Eqs. (A.3) above as the Pauli matrices, σ_a.

At an intuitive level, an algebra can be thought of as a vector space with the usual operation of vector addition, and a second binary operation (generally called the commutator or Lie bracket). In the case of the circle group, the Lie algebra is one-dimensional, and the (single) basis vector defines a direction of rotation. The entire group can be built up by infinitesimal iterations of this rotation, and hence the group is said to be generated by the basis vector of the corresponding Lie algebra. General elements of SU(2) are created (in analogy to the U(1) example $\exp(i\theta)$), by exponentiating the $\mathfrak{su}(2)$ generators multiplied by three angle parameters. Hence we have $\exp\{i\theta_a t_a\}$, where $t_a = \frac{1}{2}\sigma_a$. The commutator of the $\mathfrak{su}(2)$ generators is given by the relation $[t_a, t_b] = \mathbf{i}\epsilon_{abc}t_c$.

The groups SO(3) and SU(2) share equivalent Lie algebras, but this does not mean the two groups are identical, as a Lie algebra is a tangent space defined at the identity element. SU(2) and SO(3) are *locally* isomorphic, but globally distinct, as there are two elements in SU(2) that correspond to each element in SO(3). We say that SU(2) is a *double cover* of SO(3), and also refer to SU(2) as the *universal covering group* of SO(3).

For the purposes of explaining loop quantum gravity we will be most often concerned with SU(2), though what we are about to say can be easily generalised to other groups. The SU(2) matrices act on vectors with two complex components. Such vectors can be written as linear combinations of the two basis vectors

$$\begin{pmatrix} 1 \\ 0 \end{pmatrix}, \quad \begin{pmatrix} 0 \\ 1 \end{pmatrix}.$$

In the case of intrinsic angular momentum these correspond to pure "spin-up" and "spin-down" states. Multiplying these by $\hbar t_3 = \frac{1}{2}\hbar\sigma_3$ yields the eigenvalues $\pm\frac{1}{2}\hbar$. The set of eigenvalues $\{+\hbar/2, -\hbar/2\}$ is an example of a *multiplet* of states. As the discussion above suggests, different representations of SU(2) are possible, and these act upon different multiplets. Note that if we set $\hbar = 1$ the number of states in the multiplet is equal to $2j + 1$, where j is the largest eigenvalue in the multiplet.

Different representations can therefore be referred to by the largest eigenvalue in the corresponding multiplet. It is a common and unfortunate practice, however, to refer to the eigenvalues as a representation of the group. This can be confusing, as the eigenvalues do not have the defining property of a representation, namely that $D(g_a)D(g_b) = D(g_a \circ g_b)$ for all a, b. Therefore when the student encounters statements that the loops from which loop quantum gravity derives its name are "labelled by representations of SU(2)", it should be understood that what is really meant is that representations of SU(2) are associated to the loops and in turn the loops "are labelled by values drawn from the multiplets upon which representations of SU(2) act", a point we have also attempted to emphasise in Sect. 6.1.

A.2   Lorentz Lie-Algebra

In Sect. 4.3.3 we discussed the Lorentz group, which consists of rotations around three axes, generated by J_x, J_y, J_z, and boosts along three axes, generated by K_x, K_y, K_z. It was noted that a 2×2 representation of the rotations and boosts can be constructed from the Pauli matrices, as per Eq. (4.44). It is useful to recognise that the J_a can be regarded as purely spatial rotations, while the K_a can be thought of as rotations in the three planes defined by the time axis and each of the three spatial axes.

The generators of the n-dimensional representation of the Lorentz Lie algebra can be written in terms of the $(n \times n)$ Dirac gamma matrices $\{\gamma^I\}$, which satisfy the anticommutation relations

$$\left\{\gamma^I, \gamma^J\right\} = 2g^{IJ} \times \mathbb{1}_{n \times n} \tag{A.4}$$

where g^{IJ} is the metric tensor and $\mathbb{1}_{n \times n}$ is the $n \times n$ identity matrix.

For the case of $n = 4$, a possible choice of the matrices is given by

$$\gamma^0 = \begin{pmatrix} 0 & 1 \\ -1 & 0 \end{pmatrix}, \quad \gamma^a = \begin{pmatrix} 0 & \sigma^a \\ \sigma^a & 0 \end{pmatrix} \tag{A.5}$$

where σ^a are the usual Pauli matrices, and in this case g^{IJ} is equivalent to $\eta^{IJ} = \text{diag}(-1, 1, 1, 1)$, the usual Minkowski metric.

In terms of the $\{\gamma^\mu\}$, the generators of the Lorentz group SO(3, 1) can be written as [1]

$$T^{IJ} = \frac{i}{4}\left[\gamma^I, \gamma^J\right]. \tag{A.6}$$

Note that, whereas in the above we have restricted ourselves to the case of $3 + 1$ dimensions, the expression for the generators of the Lorentz group goes through in any dimension, with either Lorentzian or Euclidean metric [1, Sect. 3.2]. An $\mathfrak{so}(3, 1)$-valued connection can then be written as

$$\mathbf{A}_\mu = A_\mu{}^{IJ} T_{IJ} = \frac{i}{4} A_\mu{}^{IJ} [\gamma_I, \gamma_J] \tag{A.7}$$

but by the antisymmetry of the gamma matrices, the above expression can be shortened to $\mathbf{A}_\mu = \frac{i}{2} A_\mu{}^{IJ} \gamma_I \gamma_J$, where we remember that the connection is antisymmetric in the internal indices $A^{IJ} = -A^{JI}$.

Blades, Forms, and Duality

B

The notion of self-/anti-self-duality of the gauge field $F_{\alpha\beta}$ is central to understanding both the topological sector of Yang-Mills theory and the solutions of Einstein's equations in the connection formulation. As discussed in Sect. 3.2, the use of multivectors and k-forms can be very helpful for understanding duality. They also occur in the formulation of BF theory and hence play a role in setting the stage for spinfoams, as well as various approaches to writing GR as a gauge theory. Let us review these concepts.

B.1 Blades and Multivectors

A vector is normally visualised as a directed line segment with a magnitude which is interpreted as a length. One way to form the product of two vectors $\vec{u}$ and $\vec{v}$ is the dot product $\vec{u} \cdot \vec{v}$, which is a scalar that is maximised when the vectors are parallel. We can also form the wedge product, $\vec{u} \wedge \vec{v}$, which is a directed surface spanned by $\vec{u}$ and $\vec{v}$ (the direction being both an orientation in space and a preferred direction of rotation around the boundary of the surface), with a magnitude interpreted as the area of the surface. This directed surface is called a bivector, and its magnitude is maximised when $\vec{u}$ and $\vec{v}$ are perpendicular (and zero when they are parallel). The wedge product of three non-coplanar vectors is a trivector, which may be visualised as a parallelipiped with a direction (interpreted as a preferred direction assigned to a path around the edges of the parallelipiped) and a magnitude (interpreted as its volume). The wedge product of k vectors (assuming they are not parallel, coplanar, etc.) will in general be called a k-blade, and can be visualised as an oriented k-dimensional parallelipiped,[2] with a magnitude given by its enclosed

---

[2] Strictly speaking it could be any shape but a parallelipiped is most directly visualised, by imagining that each of the k vectors forming a k-blade define the length and orientation of its edges.

© Springer Nature Switzerland AG 2024
S. Bilson-Thompson, *Loop Quantum Gravity for the Bewildered*,
https://doi.org/10.1007/978-3-031-43452-5

volume.[3] A scalar may be regarded as a 0-blade. We will use u, v, w (without arrows) to denote k-blades in general, and retain $\vec{u}$, $\vec{u}$, $\vec{w}$ specifically for ordinary vectors (i.e. 1-blades).

Consider two vectors $\vec{u} = u^1 e_1 + u^2 e_2$ and $\vec{v} = v^1 e_1 + v^2 e_2$ with orthonormal basis vectors e_1, e_2 (we restrict ourselves to two dimensions for simplicity but can easily generalise). The most general product of the vectors $\vec{u}$ and $\vec{v}$, which we will refer to as their Clifford product, is

$$\vec{u}\vec{v} = (u^1 e_1 + u^2 e_2)(v^1 e_1 + v^2 e_2)$$
$$= u^1 v^1 (e_1)^2 + u^1 v^2 e_1 e_2 + u^2 v^1 e_2 e_1 + u^2 v^2 (e_2)^2 \tag{B.1}$$

i.e. we don't assume the basis vectors commute. In the special case $\vec{u} = \vec{v}$ we let $\vec{u}\vec{v} = \vec{u}\cdot\vec{u}$, hence

$$\vec{u}\vec{v} = (u^1)^2 (e_1)^2 + u^1 u^2 e_1 e_2 + u^2 u^1 e_2 e_1 + (u^2)^2 (e_2)^2$$
$$= (u^1)^2 (e_1)^2 + u^1 u^2 (e_1 e_2 + e_2 e_1) + (u^2)^2 (e_2)^2$$
$$= (u^1)^2 + (u^2)^2 .$$

In other words, since u^1 and u^2 commute (they're scalars) the terms involving $e_1 e_2$ and $e_2 e_1$ must sum to zero, so $(e_1)^2 = (e_2)^2 = 1$ while $e_1 e_2 = -e_2 e_1$. The product in Eq. (B.1) can then be written as

$$\vec{u}\vec{v} = \vec{u}\cdot\vec{v} + (u^1 v^2 - u^2 v^1)e_1 e_2 .$$

The antisymmetric term is the wedge product, $\vec{u} \wedge \vec{v} = (u^1 v^2 - u^2 v^1)e_1 e_2$. So the Clifford product of the vectors $\vec{u}$ and $\vec{v}$ is the sum

$$\vec{u}\vec{v} = \vec{u}\cdot\vec{v} + \vec{u} \wedge \vec{v} . \tag{B.2}$$

Clearly $\vec{u} \wedge \vec{v} = -\vec{v} \wedge \vec{u}$. It follows that

$$\vec{u} \cdot \vec{v} = \frac{1}{2}(\vec{u}\vec{v} + \vec{v}\vec{u}) = \frac{1}{2}\{\vec{u}, \vec{v}\} \tag{B.3}$$

$$\vec{u} \wedge \vec{v} = \frac{1}{2}(\vec{u}\vec{v} - \vec{v}\vec{u}) = \frac{1}{2}[\vec{u}, \vec{v}] \tag{B.4}$$

Considering orthonormal basis vectors e_i, e_j etc. we find that the Clifford product $e_i e_i = e_i \cdot e_i$. Conversely when $i \neq j$ we have $e_i e_j = e_i \wedge e_j$ since in this case $e_i \cdot e_j = 0$. We adopt the notation $e_i e_j = e_i \wedge e_j = e_{ij}$, and this product defines a basis bivector. Likewise $e_i e_j e_k = e_i \wedge e_j \wedge e_k = e_{ijk}$, and so forth. Since any 1-blade (i.e. vector) can be written as a linear combination of basis vectors e.g.

---

[3] The terminology "k-vector" is also sometimes used, but we avoid it here as it can cause confusion with vectors in k dimensions.

$\vec{u} = u^1 e_1 + u^2 e_2 + \dots$ it is straightforward to extend these concepts to k-blades in general, so that any 2-blade (bivector) can be written as a linear combination $v = v^{12} e_{12} + v^{23} e_{23} + \dots$ (where the v^{ij} are scalars), and so forth for 3-blades, 4-blades, etc. A linear combination of k-blades, which may involve several different values of k (e.g. $w = w^1 e_1 + w^2 e_2 + w^3 e_{12}$ which is a sum of vector and bivector parts) is referred to as a multivector, or sometimes a Clifford vector. We can readily generalise Eq. (B.2) to define the Clifford product of k-blades as

$$uv = u \cdot v + u \wedge v. \tag{B.5}$$

The importance of k-blades and multivector quantities in physics has already been touched upon in Sect. 3.2 when discussing the field-strength tensor. Their importance can be further illustrated if we consider the case of four-dimensional Minkowski spacetime, where the scalar product is taken using the metric $\eta_{\mu\nu}$. Hence $e_0 e_0 = -1$, and $e_1 e_1 = e_2 e_2 = e_3 e_3 = +1$. In this case there is an isomorphism between the basis vectors and the Dirac gamma matrices, γ_μ, and the reader can verify that the basis vectors satisfy $\{e_\mu, e_\nu\} = 2\eta_{\mu\nu}$, the defining relation of the Dirac matrices (see Eq. A.4). Since this anticommutator is formed by taking Clifford products of the e_μ, the gamma matrices are said to generate a representation of a Clifford algebra.[4]

With the scalar values of $e_i e_i$ established (for all values of i) the product uv in Eq. (B.5) can be explicitly evaluated by writing u and v in terms of the basis vectors e_i. As a result, Eq. (B.5) remains valid even when u is a k-blade and v is an l-blade with $k \neq l$, or when either or both of u and v are multivectors (i.e. linear combinations of 0-blades, 1-blades, 2-blades, etc.)

A bivector is said to be *simple* if it can be written as the wedge product of exactly two vectors. This is always possible in two or three dimensions, e.g. the sum of the two bivectors $e_{12} + e_{23} = e_1 \wedge e_2 - e_3 \wedge e_2 = (e_1 - e_3) \wedge e_2$, and hence this is the wedge product of two vectors. However it is not necessarily possible in four or more dimensions, e.g. $e_{12} + e_{34}$ is not a simple bivector.

B.2   Differential Forms and the Exterior Derivative

In Sect. 3.2 we mentioned the close correspondence between k-blades and differential forms. The most apparent distinction between these concepts is that while a 1-blade has a magnitude interpreted as a length, a 1-form has a magnitude which is interpreted as a density (with 2-blades, 2-forms, 3-blades etc. having magnitudes which generalise to area, area-density, volume etc.)

Differential forms are a tool applied to a wide variety of manifolds, but we will take some pedagogical liberties including restricting ourselves to $\mathbb{R}^N$. As a result the discussion that ensues will hopefully be easy to follow, and build up the relevant

---

[4] It should be emphasised that, as noted at the start of Sect. A.2, the Dirac gamma matrices provide a *representation* of the algebra, but as such they are not the fundamental objects under consideration. That honour belongs to the basis vectors themselves.

concepts, but may cause more experienced pure mathematicians to snarl with indignation. Much of the discussion that follows is distilled from Chaps. 4 and 5 of [2], and a more precise, detailed exposition can be found therein.

The standard example of a 1-form is the differential dx, while the equivalent example of a vector is $\frac{\partial}{\partial x}$. To understand why dx can be thought of as a density (rather than a vector of infinitesimal length, which is how it's usually introduced in elementary calculus classes) it helps to think of the directional derivative. This is the scalar product $\vec{v} \cdot \nabla f$ of the gradient of a function f with some vector $\vec{v}$, and defines how f changes as we move along the direction defined by $\vec{v}$.

The important thing to keep in mind here is that ∇f is a function which embodies all the information for how f changes in any direction. If we specify a point P (by selecting particular values of the variables upon which f depends) we are left with the information defining how much f varies under any displacement away from P. A particular displacement is defined by providing a vector $\vec{v}$. So ∇f at P is a mapping from vectors to scalars. But the mappings from vectors to scalars are just what we think of as forms, visualised as contour lines, with the number of contour lines a vector crosses defining a scalar product of the corresponding vector and form. Since the point P can vary, we can think of ∇f as a field of such mappings (or "covectors"), one for each possible choice of P.

Now consider the total derivative of a function f,

$$df = \frac{\partial f}{\partial x^1} dx^1 + \frac{\partial f}{\partial x^2} dx^2 + \ldots \tag{B.6}$$

The minimalistic change of notation from ∇f to df should serve as a clue that these are similar entities, and of course the partial derivatives occurring in Eq. (B.6) are simply the components of ∇f. It therefore makes sense to think of dx^1, dx^2 etc. as the basis 1-forms that df is composed of. We can then think of writing dx^i not to denote an infinitesimal change in the quantity x^i, but as a notation similar to $\hat{e}_i$ for unit basis vectors. The total derivative is a specific example of a more general concept which we will call the *exterior derivative*.

Consider a k-form, $\alpha = \alpha^1 \wedge \alpha^2 \wedge \ldots \alpha^k$ and an l-form $\beta = \beta^1 \wedge \beta^2 \wedge \ldots \beta^l$, where k might be equal to l, but need not be. The exterior derivative of their product will be

$$d(\alpha \wedge \beta) = d\alpha \wedge \beta + (-1)^k \alpha \wedge d\beta. \tag{B.7}$$

The second term acquires a factor of $(-1)^k$ because placing d to the right of α involves swapping a series of wedge products, picking up a factor of -1 with each swap. If we were to write a 1-form as $\alpha = \alpha_i dx^i$ application of Eq. (B.7) would yield

$$d\alpha = d\alpha_i \wedge dx^i. \tag{B.8}$$

Prompted by Eq. (B.7) the reader may anticipate a second term on the right, however we use this particular example to introduce the important relation $d(df) = 0$ obeyed by the exterior derivative, which ensures that the anticipated term vanishes. A similar line of reasoning can be applied to the case of a general k-form.

B.3   Duality

Differential forms and k-blades can be seen to correspond closely. A bivector and a 2-form both define a 2-dimensional subspace of whatever manifold they live in. A trivector and a 3-form both define a 3-dimensional subspace, etc. However as mentioned in Sect. 3.2, k-blades can be easier to visualise, as the magnitude of a k-blade is a k-dimensional volume. It can therefore often be easier to think of how the wedge products of k-forms behave by visualising them as k-blades instead. However we visualise them, it is clear that in n dimensions a k-blade (or k-form) defines not only a k-dimensional subspace, but also an $(n - k)$-dimensional subspace which is the set of directions *not* spanned by the k-blade (or k-form) under consideration. This latter subspace is said to be dual to the former. In fact the discussion in Sect. 3.2 invoked a specific example of the concept of duality, namely a mapping between bivectors in a spacetime plane (i.e. a two-dimensional plane embedded in a four-dimensional manifold) and bivectors in the plane defined by the other two spacetime directions. Thus duality is a notion that emerges naturally from the construction of the space of k-blades, and likewise from the construction of the space of differential forms, on an n-dimensional manifold M.

Consider the case of k-blades in three dimensions. The antisymmetry of the wedge product means that the unit trivector $e_{ijk} = e_i e_j e_k$ picks up a factor of -1 each time the order of any two of its factors is swapped, hence $e_{ijk} = -e_{ikj}$, etc. and so the unit trivector is a geometrical representation of the antisymmetric tensor ϵ_{ijk}. Multiplying a vector by the unit trivector yields a bivector, and multiplying a bivector by the unit trivector yields a vector (Fig. B.1). To see why, consider the familiar cross product. Any two vectors $\vec{u}, \vec{v} \in \mathbb{R}^3$ (that are not parallel to each other) span a two-dimensional subspace of $\mathbb{R}^3$. Using these two vectors we construct a third vector $\vec{w} = \vec{a} \times \vec{b}$, where the components of $\vec{w}$ are given by $w_i = \epsilon_{ijk} u_j v_k$ (and we remind the reader that summation is performed over any repeated indices, as the raising or lowering of indices is irrelevant in $\mathbb{R}^N$). This construction is taught to us in elementary algebra courses, but never quite seemed to make complete sense because it seemed to be peculiar to three dimensions. The product $\vec{u} \times \vec{v}$ is a vector which is

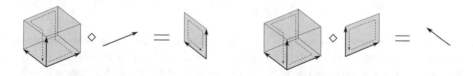

Fig. B.1 The unit trivector e_{123} allows us to explore duality in three dimensions. When we take the Clifford product, indicated here by $\diamond$, of the unit trivector with a vector, the part of e_{123} parallel to the vector yields a scalar factor via the dot product, and a factor of zero via the wedge product part. This leaves us with a bivector perpendicular to the original vector (left). Likewise the Clifford product of e_{123} with a bivector yields a vector (right). In each case the bivector and vector are dual to each other, since each spans the directions the other doesn't. Duality is therefore an extension of the concept of orthogonality. For a four-dimensional object, the dual would be taken with e_{1234}, the dual of a vector would be a trivector, and the dual of a bivector would be another bivector perpendicular to the plane defined by the vectors $\vec{u}$ and $\vec{v}$. But this plane is the same one that the wedge product $\vec{u} \wedge \vec{v}$ lies in. If we take the Clifford product of $\vec{u} \wedge \vec{v}$ with the unit 3-vector, $e_1 \wedge e_2 \wedge e_3 = e_{123}$ we are left with a vector that is perpendicular to the plane of $\vec{u} \wedge \vec{v}$, and which equals $-(\vec{u} \times \vec{v})$. Why? Because the components of $\vec{u} \wedge \vec{v}$ parallel with components of the unit trivector yield scalars, leaving only the components perpendicular to $\vec{u} \wedge \vec{v}$, as we can see by expanding the Clifford product in full,

$$
\begin{aligned}
(e_1 \wedge e_2 \wedge e_3)(\vec{u} \wedge \vec{v}) &= (e_{123})\big[(u_1 e_1 + u_2 e_2 + u_3 e_3) \wedge (v_1 e_1 + v_2 e_2 + v_3 e_3)\big] \\
&= (e_{123})\big[(u_1 v_2 - u_2 v_1)e_{12} + (u_1 v_3 - u_3 v_1)e_{13} + (u_2 v_3 - u_3 v_2)e_{23}\big] \\
&= (u_1 v_2 - u_2 v_1)e_{12312} + (u_1 v_3 - u_3 v_1)e_{12313} + (u_2 v_3 - u_3 v_2)e_{12323} \\
&= (u_1 v_2 - u_2 v_1)(-e_3) + (u_1 v_3 - u_3 v_1)e_2 + (u_2 v_3 - u_3 v_2)(-e_1) \\
&= -\vec{u} \times \vec{v}
\end{aligned}
$$

where in the third line we have dealt with the excessively-indexed terms $e_{12312} = e_1 e_2 e_3 e_1 e_2$ etc. by using the antisymmetry of the product of dissimilar terms $e_i e_j = e_i \wedge e_j = -e_j \wedge e_i = -e_j e_i$ to rearrange the basis vector terms, so that we may eliminate some of them using $e_i e_i = e_i \cdot e_i = 1$. We also find that the wedge product of $\vec{u}$ and $\vec{v}$ has components $(\vec{u} \wedge \vec{v})_{ij} = u_{[i} v_{j]}$.

This allows us to view the cross product as a three-dimensional special case of a procedure that can be performed in any number of dimensions. This procedure is "forming the dual". We can say that the cross product of two vectors in three dimensions is (up to a sign) the dual of the wedge product, and write $(\vec{u} \times \vec{v}) = \star(\vec{u} \wedge \vec{v})$.

This concept can be generalised to n dimensions. In the language of multivectors, it involves taking the Clifford product with the unit n-blade $e_1 e_2 \ldots e_n = e_{12\ldots n}$. Re-worded in the language of differential forms, any k-form $F^{a_1 a_2 \ldots a_k}$, defined on an n dimensional manifold M, can be mapped to an $(n-k)$-form $(\star F^{a_1 a_2 \ldots a_{n-k}})$ by utilising the completely antisymmetric tensor $\epsilon^{a_1 a_2 \ldots a_n}$ on M:

$$
(\star F)^{a_1 \ldots a_{n-k}} = \frac{1}{(n-k)!} \epsilon^{a_1 \ldots a_{n-k}}{}_{a_{n-k+1} \ldots a_n} F^{a_{n-k+1} \ldots a_n} . \tag{B.9}
$$

From now on we will focus on k-forms rather than k-blades. But their equivalence, and the geometric interpretation arising from this, should be kept in mind. It should be clear that in n dimensions there exists one unit 0-form,[5] n unit 1-forms, $n(n-1)$ unit 2-forms, etc. with the number of unit k-forms increasing as k approaches $n/2$ but *decreasing* as k exceeds $n/2$. A similar result is true for k-blades. We let $^n\Omega_k$

---

[5] The reader is reminded that a k-form is a scalar-valued function—i.e. a mapping that takes k vectors as arguments and returns a scalar. When we refer to a "unit" 0-form, we simply mean that we intend all scalars to be regarded as multiples of the value this form returns when provided with no vectors as arguments, so we can conveniently think of the unit 0-form itself as the scalar 1.

denote the subspace consisting only of forms of order k in n dimensions e.g. in three dimensions the space of two-forms ${}^3\Omega_2$ is spanned by the basis $\{dx^1 \wedge dx^2,$ $dx^2 \wedge dx^3,\ dx^3 \wedge dx^1\}$ where $\{x^1, x^2, x^3\}$ is some local coordinate patch—i.e. a mapping from a portion of the given manifold to a region around the origin in $\mathbb{R}^3$. In n dimensions then, the full space of differential forms is given by $\oplus_{k=0}^{n}\,{}^n\Omega_k$. One can show [2,3] that ${}^n\Omega_k = {}^n\Omega_{n-k}$, i.e. the space of k-forms is the same as the space of $(n-k)$-forms.[6]

B.4   Field Strength and the Exterior Derivative

Returning our attention to the exterior derivative, recall Eq. (B.8) and the discussion that followed, from which we recognise that the exterior derivative of a k-form is a $(k+1)$-form. This feature of the exterior derivative allows us to encapsulate several familiar differential operators in one.

The exterior derivative of a scalar (or more generally, a scalar-valued function of several variables) is a 1-form. This turns out to be equivalent to the result from elementary calculus that the gradient of a scalar-valued function is a vector. Now consider a 1-form in three dimensions, $\alpha = \alpha_1 dx^1 + \alpha_2 dx^2 \alpha_3 dx^3$. The exterior derivative of this is

$$
\begin{aligned}
d\alpha =\ & d(\alpha_1 dx^1) + d(\alpha_2 dx^2) + d(\alpha_3 dx^3)\\[4pt]
=\ & d\alpha_1 \wedge dx^1 + d\alpha_2 \wedge dx^2 + d\alpha_3 \wedge dx^3\\[4pt]
=\ & \frac{\partial \alpha_1}{\partial x^1} dx^1 \wedge dx^1 + \frac{\partial \alpha_1}{\partial x^2} dx^2 \wedge dx^1 + \frac{\partial \alpha_1}{\partial x^3} dx^3 \wedge dx^1\\[4pt]
& + \frac{\partial \alpha_2}{\partial x^1} dx^1 \wedge dx^2 + \frac{\partial \alpha_2}{\partial x^2} dx^2 \wedge dx^2 + \frac{\partial \alpha_2}{\partial x^3} dx^3 \wedge dx^2\\[4pt]
& + \frac{\partial \alpha_3}{\partial x^1} dx^1 \wedge dx^3 + \frac{\partial \alpha_3}{\partial x^2} dx^2 \wedge dx^3 + \frac{\partial \alpha_3}{\partial x^3} dx^3 \wedge dx^3\\[4pt]
=\ & \left(\frac{\partial \alpha_2}{\partial x^2} - \frac{\partial \alpha_1}{\partial x^2}\right) dx^1 \wedge dx^2\\[4pt]
& + \left(\frac{\partial \alpha_1}{\partial x^3} - \frac{\partial \alpha_3}{\partial x^1}\right) dx^3 \wedge dx^1\\[4pt]
& + \left(\frac{\partial \alpha_3}{\partial x^2} - \frac{\partial \alpha_2}{\partial x^3}\right) dx^2 \wedge dx^3\\[4pt]
\therefore\ d\alpha =\ & \partial_{[i}\alpha_{j]} dx^i \wedge dx^j
\end{aligned}
\tag{B.10}
$$

where we have used the total derivative Eq. (B.6) and the antisymmetry of the wedge product. Using the above discussion about duality to identify $dx^2 \wedge dx^3$ with $\hat{e}_1$ etc.

---

[6] In the first edition of this book the notation ${}^n\Lambda_k$ was used, however the symbol Λ^k is also used throughout the wider literature to denote the space of products of k vectors. Forms can be regarded as the elements of a vector space, so this notation is general enough to be—strictly speaking—valid. However to be specific the symbol Ω will be used here to denote the space of differential forms.

we see that when α is a 1-form, $d\alpha$ is equivalent to the curl of a 1-blade (vector) that has the same components as α.

Taking the next logical step we find that the exterior derivative of a 2-form is a 3-form. In three dimensions a 2-form can (as above) be equated to a vector, while a 3-form, proportional to $dx^1 \wedge dx^2 \wedge dx^3$ can be equated (again, via the concept of duality) to a scalar. The divergence of a vector field is a scalar, and so it should come as no surprise to find that the exterior derivative of a 2-form is equivalent to the divergence of a vector field.

Thus the exterior derivative embodies and generalises several familiar differential operators. Part of the beauty of this approach is that the exterior derivative "takes care of" deciding which differential operator to apply in a given situation, and so the use of indices to keep track of whether we're operating on a scalar, vector, or other quantity, and which component thereof, loses its importance. For example, the previously-introduced identity $d(df) = 0$ embodies the identities

$$\nabla \times (\nabla f) = 0$$
$$\nabla \cdot (\nabla \times f) = 0$$

and works appropriately to whatever type of object f happens to be.

The gradient, divergence, and curl occur specifically in the classical theory of electromagnetism. We have already seen (Sect. 3.2) how the electromagnetic field strength tensor can be written in term of bivectors (that is to say, 2-blades). We can similarly write the field strength in terms of 2-forms. To do so we construct a magnetic field 2-form, an electric field 2-form, and a field-strength 2-form which is their sum;

$$B = B_1 dx^2 \wedge dx^3 + B_2 dx^3 \wedge dx^1 + B_3 dx^1 \wedge dx^2$$
$$E = E_1 dx^1 \wedge dx^0 + E_2 dx^2 \wedge dx^0 + E_3 dx^3 \wedge dx^0$$
$$F = E + B \tag{B.11}$$

where $dx^0 = dt$. Comparing this with Eq. (3.21) the equivalence of the two formulations is clear. Applying the exterior derivative to B and writing dB_a as the total derivative

$$dB_a = \frac{\partial B_a}{\partial t} dt + \frac{\partial B_a}{\partial x^1} dx^1 + \frac{\partial B_a}{\partial x^2} dx^2 + \frac{\partial B_a}{\partial x^3} dx^3 \tag{B.12}$$

we find

$$dB = \frac{\partial B_a}{\partial t} dt \wedge dx^b \wedge dx^c + \frac{\partial B_a}{\partial x^1} dx^1 \wedge dx^b \wedge dx^c$$
$$+ \frac{\partial B_a}{\partial x^2} dx^2 \wedge dx^b \wedge dx^c + \frac{\partial B_a}{\partial x^3} dx^3 \wedge dx^b \wedge dx^c$$
$$= \frac{\partial B_a}{\partial t} dx^b \wedge dx^c \wedge dt + \frac{\partial B_a}{\partial x^a} dx^1 \wedge dx^2 \wedge dx^3 \tag{B.13}$$

where $\{a, b, c\}$ correspond to cyclic permutations of $\{1, 2, 3\}$, we have used $dx^a \wedge dx^a = 0$ to simplify between the first line and the second, and there is a sum over $a = 1, 2, 3$ in the last term on the second line. The first term on the second line is simply proportional to the time derivative of B which we will write as $\partial_t B$, while the second term is equivalent to the divergence of B, which we will write as $d_x B$ as it is the exterior derivative restricted to the spatial directions.

Similarly we find that in the exterior derivative of E the terms proportional to $(\partial E_a / \partial t) dt$ vanish, and we are left with only $d_x E$ terms which are proportional to dt and correspond with the curl of E, as per our previous result regarding the (three-dimensional) exterior derivative of a 2-form. The exterior derivative of F can therefore be written

$$dF = d_x B + \left(\frac{\partial B_a}{\partial t} dx^b \wedge dx^c \wedge dt + d_x E \right) \tag{B.14}$$

where the term in brackets is proportional to dt and the other term is not. Equation (B.14) is therefore equivalent to two separate equations, and setting

$$dF = 0 \tag{B.15}$$

requires that the divergence of B and the bracketed term in Eq. (B.14) must each be zero. Equation (B.15) is therefore equivalent to the two homogeneous Maxwell's equations.

In a similar fashion the inhomogeneous Maxwell's equations can be written as

$$\star d \star F = J \tag{B.16}$$

where the dual field strength $\star F$ is found by swapping $E_i \to -B_i$ and $B_i \to E_i$ as noted in Sect. 3.2, and the current J is an appropriately-defined 1-form.

The dual operation has another nice trick up its sleeve. Consider k-blades again, specifically a 1-blade $u = u^1 e_1 + u^2 e_2 + u^3 e_3$. The dual of this can be easily shown to take the form $\star u = u^1 e_{23} + u^2 e_{31} + u^3 e_{12}$. We keep in mind that the wedge product of any terms proportional to the same basis vector will be zero, while for orthogonal basis vectors the wedge product is equivalent to the Clifford product (e.g. $e_1 \wedge e_{31} = 0$, and $e_1 \wedge e_{23} = e_{123}$). It is then a fairly easy exercise to show that

$$u \wedge \star u = (u^1)^2 e_{123} + (u^2)^2 e_{123} + (u^3)^2 e_{123}. \tag{B.17}$$

This is clearly proportional to (in fact, the dual of) the dot product of u with itself. It is easy to visualise how this arises, even without doing the calculation explicitly. Recalling that duality is a generalisation of the concept of orthogonality, the properties of the wedge product ensure that only those terms which are dual to each other (such as e_1 and e_{23}) will be non-zero. The reader may confirm for themselves that an equivalent result is found if we replace the 1-blade u with a 1-form instead. This suggests that we can regard $u \wedge \star v$ as equivalent to the inner product of u and v.

Extending this line of reasoning, if we write the field strength $F_{\mu\nu}$ in terms of bivectors as in Eq. (3.21) and the dual field strength $\star F_{\mu\nu}$ defined in Eq. (3.22) equivalently, then we find that their wedge product $F \wedge \star F$ is equivalent to $\frac{1}{2} F_{\mu\nu} F^{\mu\nu}$. Naturally the same result applies if we write the field strength in terms of 2-forms instead of bivectors.

The discussion of classical electromagnetism is somewhat peripheral to formulating a theory of quantum gravity. It is however, for the reader unfamiliar with the language of differential forms, useful for illustrating the way in which the notation of differential forms is used, and hints at how this notation can be generalised to any number of dimensions, and arbitrary manifolds without the need to juggle a profusion of indices. It also, hopefully, lends plausibility to the use of index-free formulations of various quantities such as the actions arising in gauge field theories and formulations of GR, which will be examined further in the following sections and especially Appendix G.

B.5   Spacetime Duality

From the discussion above and in Sect. B.3, it should be apparent that in four dimensions the dual of any two-form is another two-form

$$\star F_{\alpha\beta} = \frac{1}{2}\epsilon_{\alpha\beta}{}^{\mu\nu} F_{\mu\nu} \tag{B.18}$$

(also compare this with Eq. (3.22), and as noted there, the quantity defined on the plane between any pair of spacetime axes is associated to the quantity defined on the plane between the other two spacetime axes). It is due to this property of even-dimensional manifolds that we can define *self-dual* and *anti-self-dual* k-forms, where a form is self-/anti-self-dual if

$$\star F = \pm F. \tag{B.19}$$

Given an arbitrary 2-form $G_{\mu\nu}$ its self-dual part G^+ and anti-self-dual part G^- are given by

$$G^+ = \frac{G + \star G}{2\alpha}, \qquad G^- = \frac{G - \star G}{2\beta},$$

where α and β are constants we have introduced for later convenience. We can check that

$$\star(G \pm \star G) = \pm(G \pm \star G) \tag{B.20}$$

because $\star\star = 1$ in a Euclidean background. In other words $\star G^+ = G^+$ and $\star G^- = -G^-$, which is precisely the definition of (anti-)self-duality. Thus any 2-form can always be written as a linear-sum of a self-dual and an anti-self-dual piece

$$G = \alpha G^+ + \beta G^-, \qquad \star G = \alpha G^+ - \beta G^-.$$

The above results hold for a Euclidean spacetime. For a Lorentzian background we would instead have $\star\star = -1$ and the dual of a two-form is given by

$$\star F_{\alpha\beta} = \frac{\mathbf{i}}{2}\epsilon_{\alpha\beta}{}^{\gamma\delta}F_{\gamma\delta} \tag{B.21}$$

and the statement of (anti-)self-duality becomes

$$\star F = \pm\mathbf{i}F \tag{B.22}$$

with the self-dual and anti-self-dual pieces of a two-form G being given by $G^+ = (G + \star\mathbf{i}G)/2\alpha$ and $G^- = (G - \star\mathbf{i}G)/2\beta$.

B.6 Lie-Algebra Duality

The previous section discussed self-duality in the context of tensors with spacetime indices $T^{\alpha\beta\cdots}{}_{\gamma\delta\dots}$. In gauge theories based on non-trivial Lie algebras we also have tensors with Lie algebra indices, such as the curvature $F_{\mu\nu}{}^{IJ}$ of the gauge connection $A_\mu{}^{IJ}$ where I, J label generators of the relevant Lie algebra.. The dual of the connection can then be defined using the completely antisymmetric tensor acting on the Lie algebra indices, as in

$$\star A_\mu{}^{IJ} - \frac{1}{2}\epsilon^{IJ}{}_{KL}A_\mu{}^{KL}. \tag{B.23}$$

B.7 Yang-Mills

Let us illustrate the utility of the notion of self-duality by examining the classical Yang-Mills action. We write this in a manner consistent with the discussion at the end of Sect. B.4,

$$S_{\text{YM}} = \int_{R^4} \text{Tr}\,[F \wedge \star F]$$

Varying this action with respect to the connection gives us the equations of motion[7]

$$dF = 0, \quad d\star F = 0,$$

which are satisfied if $F = \pm\star F$, i.e. if the gauge curvature is self-dual or anti-self-dual. Thus for self-/anti-self-dual solutions the Yang-Mills action reduces to

$$S_{\text{YM}}^\pm = \pm\int_{R^4} \text{Tr}\,[F \wedge F]$$

which is a topological invariant of the given manifold and is known as the *Pontryagin index*. Here the $\pm$ superscript denotes whether the field is self-dual or anti-self-dual.

---

[7] See for instance the Yang-Mills theory section of the Wikipedia article on instantons.

B.8 Geometrical Interpretation

Given any (Lie algebra valued) two-form F^I_{ab} (where $I, J, K \ldots$ are Lie algebra indices) we can obtain an element of the Lie algebra by contracting it with a member of the basis of the space of two-forms, $\{dx^i \wedge dx^j\}$ where x^i denotes the ith vector and **not** the components of a vector. The components are suppressed in the differential form notation as explained in the preceding sections. The resulting Lie algebra element is

$$\Phi^I = F^I{}_{ab} \, dx^a \wedge dx^b$$

and Φ^I is the *flux* of the field strength through the two-dimensional surface spanned by $\{dx^a, dx^b\}$.

We can also define

$$\star\Phi^I = \star F^I{}_{ab} \, dx^a \wedge dx^b = \frac{1}{2}\epsilon_{ab}{}^{cd} F^I{}_{cd} \, dx^a \wedge dx^b$$

which implies that $\star\Phi^I{}_{ab} = \frac{1}{2}\epsilon_{ab}{}^{cd}\Phi^I{}_{cd}$, i.e. the flux of the field strength through the a-b plane is equal to the flux of the *dual* field through the c-d plane.

B.9 (Anti) Self-dual Connections

When we say that the connection is (anti-)self-dual, explicitly this means that

$$A^{IJ}_\mu = \pm\star A_\mu{}^{IJ} = \pm\frac{\mathbf{i}}{2}\epsilon^{IJ}{}_{KL}A_\mu{}^{KL}. \tag{B.24}$$

Let us now show the relation between the (anti-)self-dual four-dimensional connection and its restriction to the spatial hypersurface Σ. We begin by writing the full connection in terms of the generators $\{\gamma^I\}$ of the Lorentz Lie algebra, $^\pm A := A^{IJ}_\mu \gamma_I \gamma_J$, and expanding the sum (see [4, Sect. 2] and Sect. A.2), thus

$$A^{IJ}_\mu \gamma_I \gamma_J = A^{i0}_\mu \gamma_i \gamma_0 + A^{0i}_\mu \gamma_0 \gamma_i + A^{ij}_\mu \gamma_i \gamma_j$$
$$= 2A^{0i}_\mu \gamma_0 \gamma_i + A^{ij}_\mu \gamma_i \gamma_j$$
$$= 2A^{0i}_\mu \begin{pmatrix} \sigma_i & 0 \\ 0 & -\sigma_i \end{pmatrix} + \mathbf{i}A^{jk}_\mu \epsilon^{ijk} \begin{pmatrix} \sigma_i & 0 \\ 0 & \sigma_i \end{pmatrix}. \tag{B.25}$$

In the second line we have used the fact that A^{IJ}_μ is antisymmetric in the internal indices and that the gamma matrices anticommute. In the third we have used the expressions for the gamma matrices given in Sect. A.2 to expand out the matrix products. This allows us to write the last line in the above expression in the form

$$A = A^{IJ}_\mu \gamma_I \gamma_J = 2\mathbf{i} \begin{pmatrix} A^{i+}_\mu \sigma_i & 0 \\ 0 & A^{i-}_\mu \sigma_i \end{pmatrix} \tag{B.26}$$

where

$$A_\mu^{i+} = \frac{1}{2}\epsilon^{ijk}A_\mu^{jk} - \mathbf{i}A_\mu^{0i} \tag{B.27a}$$

$$A_\mu^{i-} = \frac{1}{2}\epsilon^{ijk}A_\mu^{jk} + \mathbf{i}A_\mu^{0i}. \tag{B.27b}$$

For $I = 0$, $J \in \{1, 2, 3\}$, using the definition of the dual connection, we find that

$$A_\mu{}^{0i} = \frac{\mathbf{i}}{2}\epsilon^{0i}{}_{jk}A_\mu{}^{jk}$$

and so we may rewrite these expressions as

$$A_\mu^{i+} = \frac{1}{2}\left(\epsilon^{ijk} + \epsilon^{0i}{}_{jk}\right)A_\mu^{jk} \tag{B.28a}$$

$$A_\mu^{i-} = \frac{1}{2}\left(\epsilon^{ijk} - \epsilon^{0i}{}_{jk}\right)A_\mu^{jk}. \tag{B.28b}$$

Path Ordered Exponential

From Eq. (3.26) we see that the effect of a holonomy of a connection along a path λ (for either an open or closed path) in a manifold M is defined as

$$\psi_{|(\tau=1)} = \mathcal{P}\left\{e^{\int_\lambda igd\tau' A_\mu n^\mu}\right\}\psi_{|(\tau=0)} = U_\lambda\,\psi_{|(\tau=0)}. \tag{C.1}$$

The exponential can be formally expressed in terms of a Taylor series expansion

$$e^{-\int_\lambda d\tau' A_\mu n^\mu} = 1 + \sum_{n=1}^{\infty}\frac{1}{n!}\left\{\int_{\sigma_0=0}^{\sigma_1}\int_0^{\sigma_2}\cdots\int_0^{\sigma_n=1} d\tau_1 d\tau_2\ldots d\tau_n\ A(\sigma_n)A(\sigma_{n-1})\ldots A(\sigma_1)\right\} \tag{C.2}$$

where for the nth term in the sum, the path λ is broken up into n intervals parametrized by the variables $\{\tau_1, \tau_2, \ldots, \tau_n\}$ over which the integrals are performed. The path ordering enforces the condition that the effect of traversing each interval is applied in the order that the intervals occur. The interested reader is referred to pages 66–68 of [5].

© Springer Nature Switzerland AG 2024
S. Bilson-Thompson, *Loop Quantum Gravity for the Bewildered*,
https://doi.org/10.1007/978-3-031-43452-5

ADM Variables

D

One would like to be able to determine the data required to embed the spatial hyper-surfaces Σ within the 4-manifold $\mathcal{M}$, given the spacetime metric g_{ab} and the unit time-like vector field n^a normal to Σ. This data consists of the intrinsic and extrinsic curvature tensors (h_{ab}, k_{ab}). As explained in the main text the object h_{ab} defined by Eq. (4.16) plays the role of the intrinsic metric (or "curvature") of Σ. The quantity k_{ab} is the *extrinsic curvature* of Σ determined by the particular form of its embedding in $\mathcal{M}$. In order to define k_{ab} we first need to determine the form of the covariant spatial derivative.

D.1 Covariant Spatial Derivative

To help visualize the covariant spatial derivative D_a, one can think of an arbitrary configuration of the electric field $\boldsymbol{E}$ in three-dimensional space $^3\Sigma$. For simplicity, if $^3\Sigma$ is $\mathbb{R}^3$ and $^2\Sigma \subset {}^3\Sigma$ is the surface $z = 0$, then the three-dimensional derivative operator $\boldsymbol{\nabla} = (\partial_x, \partial_y, \partial_z)$ on $\mathbb{R}^3$ reduces to the two-dimensional derivative $\boldsymbol{D} = (\partial_x, \partial_y)$ on the x-y plane. $D_a E_b$ tells us how $\boldsymbol{E}$ changes as we move from one point to another in $^2\Sigma$.

The covariant spatial derivative on Σ acting on an arbitrary spacetime tensor $T_{b_1 \ldots b_i}{}^{c_1 \ldots c_j}$ is given by [6, Sect. 3.2.2.2]

$$D_a T_{b_1 \ldots b_i}{}^{c_1 \ldots c_j} = h_a^{a'} h_{b_1}{}^{b'_1} \ldots h_{b_i}{}^{b'_i} h^{c_1}{}_{c'_1} \ldots h^{c_j}{}_{c'_j} \nabla_{a'} T_{b'_1 \ldots b'_i}{}^{c'_1 \ldots c'_j}. \tag{D.1}$$

This expression simplifies considerably in the case of a vector field n_a. Using Eq. (4.16) and the fact that $n^d \nabla_c n_d = (1/2) \nabla_c (n_d n^d) = (1/2) \nabla_c (-1) = 0$ because $n^a n_a = -1$, the spatial derivative of an arbitrary vector field can be written as

$$D_a n_b = h_a{}^c h_b{}^d \nabla_c n_d = (g_b{}^d + n_b n^d) h_a{}^c \nabla_c n_d = h_a{}^c \nabla_c n_b. \tag{D.2}$$

© Springer Nature Switzerland AG 2024
S. Bilson-Thompson, *Loop Quantum Gravity for the Bewildered*,
https://doi.org/10.1007/978-3-031-43452-5

There is nothing mysterious about (D.2). As shown by the electric field example above, it simply measures how the vector field n^a changes from point to point as we move around the spatial manifold Σ.

D.2 Extrinsic Curvature

The extrinsic curvature of a given manifold is a mathematical measure of the manner in which it is *embedded* in a manifold of higher dimension. As illustrated in Fig. 4.3, a two-dimensional cylinder embedded in $\mathbb{R}^3$ has zero intrinsic curvature, but non-zero extrinsic curvature. The normal at each point of the cylinder is a three-dimensional vector n_b and this vector changes as one moves around the cylindrical surface if the extrinsic curvature of the surface is non-zero. Thus, the simplest definition for a tensorial quantity which measures this change is given by

$$k_{ab} = D_a n_b = h_a{}^c h_b{}^d \nabla_c n_d \tag{D.3}$$

where D_a is the covariant spatial derivative defined in Sect. D.1. This quantity turns out to be symmetric. In order to see this ([6, Sect. 3.2.2.2]), note that given two spatial vector fields Y^a and Z^a, their commutator $[Y, Z]^a = Y^b \nabla_b Z^a - Z^b \nabla_b Y^a$ will also be spatial, i.e. $n_a [Y, Z]^a = 0$. Since $n_a Y^a = 0$, by applying the product rule to $\nabla_b n_a Y^a = 0$ it follows that $n_a \nabla_b Y^a = -Y^a \nabla_b n_a$. The equivalent result holds if we replace Y^a by Z^a. Substituting in these results we find that

$$n_a[Y, Z]^a = n_a (Y^b \nabla_b Z^a - Z^b \nabla_b Y^a) = -Z^a Y^b \nabla_b n_a + Z^b Y^a \nabla_b n_a$$
$$= Y^a Z^b (\nabla_b n_a - \nabla_a n_b)$$
$$= 0$$

where we have used the summation over indices to swap the labels a, b in one of the terms. Since Y^a, Z^a are purely spatial, this implies that (the spatial projection of) $\nabla_a n_b = \nabla_b n_a$. Thus the extrinsic curvature of $^3\Sigma$ can be written as

$$k_{ab} = \frac{1}{2}(D_a n_b + D_b n_a) \tag{D.4}$$

verifying the symmetry of k_{ab} which was stated, without proof, in Sect. 4.2.

D.3 Canonical Momentum in the ADM Formulation

Recall that the time vector field is written in terms of the lapse N, shift N^μ and the normal to the hypersurface n^μ, so that $t^\mu = N n^\mu + N^\mu$ (Eq. 4.11). We wish to write down the explicit form of the Lie derivative of a one-index X_a and two-index object h_{ab}, with respect to a vector field v_a. Conveniently this is already present in Eqs. (E.2)–(E.4)! As we may expect, when a vector field is a sum of two (or more) vector fields (as for the time-evolution field above), the Lie derivative with respect to that field decomposes into the sum of Lie derivatives with respect to each of the components fields. So if $X_a = u_a + v_a + w_a$, then $\pounds_X[T] = \pounds_u[T] + \pounds_v[T] + \pounds_w[T]$, where T is the arbitrary tensor whose Lie derivative we want to find. You can see this directly from Eq. (E.1) by writing the field X as a sum of other vector fields. When T is a vector, then $\pounds_X T = [X, T] = [u, T] + [v, T] + [w, T]$ and so on ($[A, B]$ is the commutator of two vector fields as in Eq. E.2).

There are two steps involved in deriving the form of the canonical momentum. First is to prove the identity (4.25). The second is to use that result to perform the functional derivative of the Einstein-Hilbert Lagrangian L_{EH} with respect to the $\dot{h}_{ab}$ to obtain Eq. (4.26).

First, we wish to show that $\pounds_{\vec{t}} h_{ab} = 2N k_{ab} + \pounds_{\vec{N}} h_{ab}$, which we can do by finding a suitable expression for $\pounds_{\vec{t}} h_{\mu\nu}$, and then restricting the indices to the range $\mu, \nu \to a, b \in \{1, 2, 3\}$. So, since $t_\mu = N n_\mu + N_\mu$, using the above mentioned additive property of Lie derivatives, we have $\pounds_{\vec{t}} h_{\mu\nu} = \pounds_{N\vec{n}} h_{\mu\nu} + \pounds_{\vec{N}} h_{\mu\nu}$. The second term is present (with indices restricted, as just discussed) in Eq. (4.25). It now remains to be shown that $2N k_{ab} = \pounds_{N\vec{n}} h_{ab}$.

The simplest approach is to recognise that the Lie derivative of a *metric tensor* with respect to the vector field $\vec{n}$ (we shall neglect the factor of N at first, but re-introduce it shortly) is given by Eq. (E.4), which we restate for convenience:

$$\pounds_{\vec{n}} h_{\mu\nu} = \nabla_\mu n_\nu + \nabla_\nu n_\mu.$$

The above equation holds true only when the derivative operator ∇_μ is compatible with the metric $h_{\mu\nu}$ whose Lie derivative we wish to determine (that is, $\nabla_\mu h_{\mu\nu} = 0$). Hence we restrict ourselves to the spatial components and switch to the correct notation D for the spatial derivative operator instead of ∇. Then by definition (D.4) we see that the Lie derivative of the spatial metric is twice the extrinsic curvature of $^3\Sigma$,

$$\pounds_{\vec{n}} h_{ab} = D_a n_b + D_b n_a = 2k_{ab}. \tag{D.5}$$

Now we equate this expression with the definition of the Lie derivative of a rank-2 tensor, Eq. (E.3), and follow the treatment of [6, Sect. 3.2.2.2]:

$$
\begin{aligned}
2k_{ab} = \pounds_{\vec{n}} h_{ab} &= n^c \nabla_c h_{ab} + h_{ac} \nabla_b n^c + h_{bc} \nabla_a n^c \\
&= \frac{1}{N} \left(N n^c \nabla_c h_{ab} + N h_{ac} \nabla_b n^c + N h_{bc} \nabla_a n^c \right) \\
&= \frac{1}{N} \left(N n^c \nabla_c h_{ab} + h_{ac} \nabla_b (N n^c) + h_{bc} \nabla_a (N n^c) \right) \\
&= \frac{1}{N} \pounds_{\vec{t} - \vec{N}} h_{ab} \\
&= \frac{1}{N} \left(\pounds_{\vec{t}} h_{ab} - \pounds_{\vec{N}} h_{ab} \right) \\
&= \frac{1}{N} h_a{}^c h_b{}^d \left(\pounds_{\vec{t}} h_{cd} - \pounds_{\vec{N}} h_{cd} \right) \\
&= \frac{1}{N} \left(\dot{h}_{ab} - D_a N_b - D_b N_a \right) \tag{D.6}
\end{aligned}
$$

where in the second line we have multiplied and divided by the scale factor N. In the third line we have used the fact that $n^c h_{ac} = 0$ to move N inside the derivative operator. In going from the third to the fourth, we have used Eq. (E.3) in reverse, along with the relationship between the lapse, shift and time-evolution fields, $Nn^a = t^a - N^a$. The fifth line is obtained by using the linearity of the Lie derivative. At this point we may readily rearrange the expression to obtain Eq. (4.25). In the sixth we have, in the words of Bojowald, "smuggled in" two factors of h knowing that k_{ab} is spatial to begin with. In the seventh, the spatial projection $h_a{}^c h_b{}^d £_t h_{cd} = \dot{h}_{ab}$ is identified as the "time-derivative" of the spatial metric. We leave the remaining step (to show that $h_a{}^c h_b{}^d £_{\vec{N}} h_{cd} = D_a N_b + D_b N_a$) as an exercise for the reader.

To summarize, we have

$$k_{ab} = \frac{1}{2N} \left[£_{\vec{t}} h_{ab} - D_{(a} N_{b)} \right] = \frac{1}{2N} \left[\dot{h}_{ab} - D_{(a} N_{b)} \right]. \tag{D.7}$$

Now, the Einstein-Hilbert Lagrangian is given by

$$L_{\text{EH}} = N\sqrt{h} \left[{}^{(3)}R + k^{ab} k_{ab} - k^2 \right].$$

The first term does not contain any dependence on k_{ab} or N_a and so its derivative with respect to $\dot{h}_{ab}$ vanishes. For the remaining two terms we have

$$\frac{\delta L_{\text{EH}}}{\delta \dot{h}_{ef}} = N\sqrt{h} \left[k^{ab} \frac{\delta k_{ab}}{\delta \dot{h}_{ef}} + k_{ab} \frac{\delta k^{ab}}{\delta \dot{h}_{ef}} - 2k \frac{\delta k}{\delta \dot{h}_{ef}} \right],$$

where $k = h^{ab} k_{ab}$ and k^{ab} can be written as $h^{ac} h^{bd} k_{cd}$, hence

$$\frac{\delta L_{\text{EH}}}{\delta \dot{h}_{ef}} = N\sqrt{h} \left[k^{ab} \frac{\delta k_{ab}}{\delta \dot{h}_{ef}} + k_{ab} h^{ac} h^{bd} \frac{\delta k_{cd}}{\delta \dot{h}_{ef}} - 2k \, h^{ab} \frac{\delta k_{ab}}{\delta \dot{h}_{ef}} \right]. \tag{D.8}$$

From (D.7) we have

$$\frac{\delta k_{ab}}{\delta \dot{h}_{ef}} = \frac{1}{2N} \delta_a^e \delta_b^f. \tag{D.9}$$

Inserting this into the previous expression we have

$$\frac{\delta L_{\text{EH}}}{\delta \dot{h}_{ef}} = N\sqrt{h} \left[k^{ab} \frac{1}{2N} \delta_a^e \delta_b^f + k_{ab} h^{ac} h^{bd} \frac{1}{2N} \delta_c^e \delta_d^f - 2k \, h^{ab} \frac{1}{2N} \delta_a^e \delta_b^f \right]$$

$$= \sqrt{h} \left[k^{ef} - k h^{ef} \right] = \pi^{ef} \tag{D.10}$$

which is identical to (4.26) as desired.

Lie Derivative

E

The Lie derivative $\pounds_X$ of a tensor T is the change in T evaluated along the flow generated by the vector field $\vec{X}$ on a manifold. When T is simply a function $T \equiv f$ on the manifold, the Lie derivative reduces to the directional derivative of f along X

$$\pounds_X T \equiv X^a \partial_a f = \frac{\partial}{\partial s} f(s) ,$$

where s parametrises the points along the curve generated by X. This fact is related to the interpretation of the differential dx as a component of a 1-form, and the derivative operator $\partial_x = \partial/\partial x$ as a component of a vector field (see Sect. B.2 for more detail if this interpretation is unfamiliar.) When the connection is torsion-free, we may replace ∂_α with ∇_α.

It can be shown [3] that

$$\pounds_X T^{\mu_1 \ldots \mu_n}_{\nu_1 \ldots \nu_m} = X^\alpha \nabla_\alpha T^{\mu_1 \ldots \mu_n}_{\nu_1 \ldots \nu_m} - \sum_{i=1}^{n} T^{\ldots \alpha \ldots}_{\nu_1 \ldots \nu_m} \nabla_\alpha X^{\mu_i} + \sum_{i=1}^{m} T^{\mu_1 \ldots \mu_n}_{\ldots \alpha \ldots} \nabla_{\nu_i} X^\alpha \quad (E.1)$$

where $\ldots \alpha \ldots$ is shorthand for an expression with α in the ith position and some number of μs or νs elsewhere, e.g. $\mu_1 \ldots \mu_{i-1} \alpha \mu_{i+1} \ldots \mu_n$. In particular the Lie derivative of a vector field T^μ along a vector field X^ν reduces to the commutator of the two vector fields,

$$\pounds_X T^\mu = X^\alpha \nabla_\alpha T^\mu - T^\alpha \nabla_\alpha X^\mu \equiv [X, T] . \quad (E.2)$$

In the case of a rank-2 tensor $T^{\mu\nu}$

$$\pounds_X T_{\mu\nu} = X^\alpha \nabla_\alpha T_{\mu\nu} + T_{\alpha\nu} \nabla_\mu X^\alpha + T_{\mu\alpha} \nabla_\nu X^\alpha . \quad (E.3)$$

© Springer Nature Switzerland AG 2024
S. Bilson-Thompson, *Loop Quantum Gravity for the Bewildered*,
https://doi.org/10.1007/978-3-031-43452-5

Applying this to the metric tensor $g_{\mu\nu}$ we find the relation

$$\pounds_X g_{\mu\nu} = \nabla_\mu X_\nu + \nabla_\nu X_\mu \tag{E.4}$$

since the covariant derivative of the metric vanishes.

3+1 Decomposition of the Palatini Action

F

Let us recall the gravity action (4.65) with connection and tetrad variables as basic variables,

$$S_P[e, \omega] = \frac{1}{4\kappa} \int d^4x \, \epsilon^{\mu\nu\alpha\beta} \epsilon_{IJKL} \, e_\mu{}^I e_\nu{}^J F_{\alpha\beta}{}^{KL}$$

where, as before $F_{\alpha\beta}{}^{KL}$ is the curvature of the gauge connection as given by (4.66):

$$F^{KL}{}_{\gamma\delta} = \partial_{[\gamma}\omega_{\delta]}{}^{KL} + \frac{1}{2}\left[\omega_\gamma{}^{KM}, \omega_{\delta M}{}^L\right].$$

As in Sect. 4.2, we assume that our spacetime manifold $\mathcal{M}$ is topologically $\Sigma_t \times \mathbb{R}$, where Σ_t are spatial (3D) manifolds which "foliate" $\mathcal{M}$. We identify a vector field $t^\mu = Nn^\mu + N^\mu$ as the generator of "time-evolution", written in terms of the purely time-like normal vector n^μ at each point of Σ_t, the lapse function N and the purely spatial "shift" vector-field N^μ.

In the metric formalism, we started by writing the 4-metric $g_{\mu\nu}$ in terms of a spatial component $h_{\mu\nu}$[8] and a time-like component $n_\mu n_\nu$ such that $g_{\mu\nu} = h_\mu h_\nu - n_\mu n_\nu$. In contrast, here we don't have a metric! Instead we have the volume form $\epsilon^{\mu\nu\alpha\beta}$ and the tetrad field $e_\mu{}^I$.

We proceed by noting that, firstly, on a p-dimensional manifold, the space of p-forms is one-dimensional, as per Sect. B.3. In other words any four-form $L^{\mu\nu\alpha\beta}$ defined on $\mathcal{M}$ is proportional to any other four-form on $\mathcal{M}$. Second, the wedge or antisymmetric outer product of a three-form and a one-form gives a four-form. Now, on the 3-manifold Σ_t, there exists a volume three-form ϵ^{abc} or $\epsilon^{\alpha\beta\gamma}$ using four-dimensional indices. One can take the wedge product of $\epsilon^{\alpha\beta\gamma}$ with the one-form t^δ

---

[8] Recalling once again that even though μ, ν are four-dimensional indices, $h_{\mu\nu}$ itself is purely spatial, because it satisfies $h_{\mu\nu}n^\nu = 0$.

© Springer Nature Switzerland AG 2024
S. Bilson-Thompson, *Loop Quantum Gravity for the Bewildered*,
https://doi.org/10.1007/978-3-031-43452-5

(the dual of the time-evolution vector field) to obtain a four-form, $\epsilon^{[\alpha\beta\gamma}t^{\delta]}$. Then by virtue of the fact that the space of four-forms on $\mathcal{M}$ is one-dimensional it follows that

$$\epsilon^{\mu\nu\alpha\beta} \propto \epsilon^{[\mu\nu\alpha}t^{\beta]}.$$

The constant of proportionality can easily be determined by contracting both sides with t_β, to obtain

$$\epsilon^{\mu\nu\alpha\beta} = 4\epsilon^{[\mu\nu\alpha}t^{\beta]}. \tag{F.1}$$

Substituting this expression into the tetrad Palatini action we find

$$
\begin{aligned}
S_P[e,\omega] &= \frac{1}{4\kappa}\int d^4x\, 4\,\epsilon^{[\mu\nu\alpha}t^{\beta]}\epsilon_{IJKL}\,e_\mu{}^I e_\nu{}^J F_{\alpha\beta}{}^{KL} \\
&= \frac{1}{4\kappa}\int d^4x\, \left(\epsilon^{\mu\nu\alpha}t^\beta + \epsilon^{\beta\mu\nu}t^\alpha + \epsilon^{\alpha\beta\mu}t^\nu + \epsilon^{\nu\alpha\beta}t^\mu\right)\epsilon_{IJKL}e_\mu{}^I e_\nu{}^J F_{\alpha\beta}{}^{KL}.
\end{aligned}
\tag{F.2}
$$

To proceed further, we introduce the following notation [7, Sect. 6.2]:

$$E^\alpha{}_{IJ} := \frac{1}{2}\epsilon^{\alpha\mu\nu}\epsilon_{IJKL}e_\mu{}^K e_\nu{}^L \tag{F.3a}$$

$$(e\cdot t)^I := t^\mu e_\mu{}^I \tag{F.3b}$$

$$(A\cdot t)^{IJ} := t^\mu A_\mu{}^{IJ}. \tag{F.3c}$$

We will also need the identity [7, Sect. 3.2]

$$t^\mu F_{\mu\nu}{}^{IJ} = \pounds_{\vec{t}}\, A_\nu{}^{IJ} - D_\nu(A\cdot t)^{IJ}. \tag{F.4}$$

Using (F.3) and (F.4), (F.2) becomes

$$
\begin{aligned}
S_P[e,\omega] = \frac{1}{\kappa}\int d^4x\, \Big[&\frac{1}{4}(e\cdot t)^I \epsilon_{IJKL}\epsilon^{\nu\alpha\beta}e_\nu{}^J F_{\alpha\beta}{}^{IJ} \\
&+\frac{1}{2}E^\alpha{}_{IJ}\pounds_{\vec{t}}A_\alpha{}^{IJ} - \frac{1}{2}E^\alpha{}_{IJ}D_\alpha(A\cdot t)^{IJ}\Big].
\end{aligned}
\tag{F.5}
$$

Substituting the expression for t^μ in terms of the lapse and shift, the first term in the above equation can be written as

$$\frac{1}{4}(e\cdot t)^I \epsilon_{IJKL}\epsilon^{\nu\alpha\beta}e_\nu{}^J F_{\alpha\beta}{}^{IJ} = -\frac{1}{2\sqrt{h}}N\mathrm{Tr}(\tilde{E}^\alpha\tilde{E}^\beta F_{\alpha\beta}) + \frac{1}{2}N^\beta\mathrm{Tr}(\tilde{E}^\alpha F_{\alpha\beta}).$$

Inserting this in the previous expression (F.5), we see that the Lagrangian of $3 + 1$ Palatini theory can be written as

$$L_P = \int_\Sigma -\frac{1}{2\sqrt{h}} N \text{Tr}(\tilde{E}^\alpha \tilde{E}^\beta F_{\alpha\beta}) + \frac{1}{2} N^\beta \text{Tr}(\tilde{E}^\alpha F_{\alpha\beta})$$

$$+ \frac{1}{2} E^\alpha{}_{IJ} \pounds_{\vec{t}} A_\alpha{}^{IJ} + \frac{1}{2} D_\alpha(E^\alpha{}_{IJ})(A \cdot t)^{IJ} \qquad \text{(F.6)}$$

where we have performed an integration by parts on the last term in (F.5) and dropped a surface term, which is presumed to vanish at spatial infinity, in (F.6).

Since our configuration variable is the connection $A_\alpha{}^{IJ}$, the canonical momentum can be read off the coefficient multiplying the time-derivative (or in this case the Lie derivative with respect to the time-evolution vector field) of the connection in the Palatini Lagrangian. Thus the canonical momentum is the triad field $E^\alpha{}_{IJ}$.

The configuration variables of our theory are N, N^α, $(A \cdot t)$, $A_\alpha{}^{IJ}$, and $E^\alpha{}_{IJ}$. However, since the Lagrangian (F.6), does not contain any time-derivatives of N, N^α and $(A \cdot t)$, these variables act as Lagrange multipliers and their respective coefficients must therefore be constant on the physical phase space.

To obtain the Hamiltonian, we can perform the usual Legendre transform on (F.6) to obtain

$$H_P = \int_\Sigma \frac{1}{2\sqrt{h}} N \text{Tr}(\tilde{E}^\alpha \tilde{E}^\beta F_{\alpha\beta}) - \frac{1}{2} N^\beta \text{Tr}(\tilde{E}^\alpha F_{\alpha\beta})$$

$$- \frac{1}{2} D_\alpha(E^\alpha{}_{IJ})(A \cdot t)^{IJ} + \lambda_{\alpha\beta} \epsilon^{IJKL} \tilde{E}^\alpha{}_{IJ} \tilde{E}^\beta{}_{KL} \qquad \text{(F.7)}$$

where the last term has been inserted in order to satisfy the constraint that

$$\epsilon^{IJKL} \tilde{E}^\alpha{}_{IJ} \tilde{E}^\beta{}_{KL} = 0$$

which follows from the definition of $E^\alpha{}_{IJ}$.

The Hamiltonian thus becomes a sum of constraints, specifically

$$\text{Tr}(\tilde{E}^\alpha \tilde{E}^\beta F_{\alpha\beta}) \approx 0, \quad \text{Tr}(\tilde{E}^\alpha F_{\alpha\beta}) \approx 0, \quad D_\alpha(E^\alpha{}_{IJ}) \approx 0$$

which are, respectively, the Hamiltonian, diffeomorphism and Gauss constraints.

The Kodama State

G

The Kodama state is an exact solution of the Hamiltonian constraint for LQG with positive cosmological constant $\Lambda > 0$ and hence is of great importance for the theory. It is given by

$$\Psi_K(A) = \mathcal{N} e^{\int S_{CS}} \tag{G.1}$$

where $\mathcal{N}$ is a normalization constant. The action $S_{CS}[A]$ is the Chern-Simons action for the connection A^I_μ on the spatial 3-manifold M, given by

$$S_{CS} = \frac{2}{3\Lambda} \int Y_{CS}$$

where

$$Y_{CS} = \frac{1}{2}\text{Tr}\left[A \wedge dA + \frac{2}{3} A \wedge A \wedge A \right]$$

with $dA \simeq \partial_{[\mu} A^I_{\nu]}$ being the exterior derivative of A. Consistent with our discussion of bivectors and k-forms in Appendix B the wedge product between two 1-forms P and Q is

$$P \wedge Q \simeq P_{[a} Q_{b]}.$$

For identical one-forms the wedge product gives zero. That is why for the Chern-Simons action to have a non-zero cubic term the connection must be non-abelian. As was the case in Sect. 5.1 the restriction to the spatial 3-manifold means we replace tetrads with triads, and indicate this by changing internal indices $I, J, K \rightarrow i, j, k \in \{1, 2, 3\}$. Let us write the various terms in the Chern-Simons density explicitly;

$$A \wedge dA \equiv A^i_{[p} \partial_q A^j_{r]} T_i T_j, \qquad A \wedge A \wedge A \equiv A^i_{[p} A^j_q A^k_{r]} T_i T_j T_k$$

© Springer Nature Switzerland AG 2024
S. Bilson-Thompson, *Loop Quantum Gravity for the Bewildered*,
https://doi.org/10.1007/978-3-031-43452-5

where $p, q, r \ldots$ are worldvolume ("spacetime") indices, $i, j, k \ldots$ are worldsheet ("internal") indices and T_i are the basis vectors of the Lie algebra/internal space. Taking the trace over these terms gives us

$$Y_{CS} = \frac{1}{2} \text{Tr} \left[A^i_{[p} \partial_q A^j_{r]} T_i T_j + \frac{2}{3} A^i_{[p} A^j_q A^k_{r]} T_i T_j T_k \right]$$

The trace over the Lie algebra elements gives us

$$\text{Tr}\left[T_i T_j\right] = \delta_{ij}, \qquad \text{Tr}\left[T_i T_j T_k\right] = f_{ijk}$$

where f_{ijk} are the structure constants of the gauge group.

Now, the Hamiltonian constraint in Ashtekar variables has the form

$$\mathcal{H} = \epsilon^{ij}{}_k \tilde{E}^a_i \tilde{E}^b_j F^k{}_{ab}.$$

We can quantize this expression in the usual way setting the connection A^i_a as the "position" and the triad $\tilde{E}^a_i$ as its conjugate "momentum". Then in terms of operators, the action of the $\hat{A}^i_a$ on a state will correspond to multiplication and $\hat{E}^a_i$ corresponds to taking the functional derivative with respect to the the connection, hence

$$\hat{A}^i_a \Psi(A) \equiv A^i_a \Psi(A), \qquad \hat{E}^a_i \Psi(A) \equiv \frac{\delta}{\delta A^i_a} \Psi(A). \tag{G.2}$$

The operator form of the Hamiltonian constraint then becomes

$$\hat{\mathcal{H}} = \hat{F}^k{}_{ab} \hat{E}^a_i \hat{E}^b_j \tag{G.3}$$

and its action on a wavefunction $\Psi(A)$ becomes

$$\hat{\mathcal{H}}\Psi(A) = \hat{F}^k{}_{ab} \frac{\delta}{\delta A^i_a} \frac{\delta}{\delta A^j_b} \Psi(A). \tag{G.4}$$

Now, making use of the fact that

$$\frac{\delta}{\delta A^i_a} \frac{\delta}{\delta A^j_b} S_{CS} = \epsilon^{ij}{}_k F^k_{ab} S_{CS}$$

we can immediately see that the Kodama state $\Psi_K(A)$ is an eigenstate of (G.4) with eigenvalue $\frac{2}{3\Lambda}$!

Despite the remarkable fact that the Ashtekar variables allow us to find an *exact* solution of the gravitational Hamiltonian for *arbitrary* geometries, there are several technical problems with treating the Kodama state as a valid wavefunction for quantum gravity as was first pointed out by Witten [8]. In recent work, Randono [9–11] has suggested that these problems can be addressed by working with a suitable modification of the original Kodama state.

Peter-Weyl Theorem

The crucial step involved in going from graph states with edges labelled by holonomies to graph states with edges labelled by group representations (angular momenta) is the Peter-Weyl theorem. This theorem allows the generalization of the notion of Fourier transforms to functions defined on a group manifold for compact, semi-simple Lie groups.

Given a group $\mathcal{G}$, let $D^j(g)_{mn}$ be the matrix representation of any group element $g \in \mathcal{G}$. Then we have (see Chap. 8 of [12]).

Theorem H.1 *The irreducible representation matrices $D^j(g)$ for the group SU(2) satisfy the following orthonormality condition*

$$\int d\mu(g) D^{\dagger}_{j}(g)^m{}_n D^{j'}(g)^{n'}{}_{m'} = \frac{n_G}{n_j} \delta^{j'}{}_j \delta^{n'}{}_n \delta^{m'}{}_m .$$ (H.1)

Here n_j is the dimensionality of the jth representation of G and n_G is the *order* of the group. For a finite group this is simply the number of elements of the group. For example, for $\mathbb{Z}_2$, $n_G = 2$. However a continuous or Lie group such as $SU(2)$ has an uncountable infinity of group elements. In such cases n_G corresponds to the "volume" of the group manifold.

This property allows us to decompose any square-integrable function $f(g) : \mathcal{G} \to \mathbb{C}$ in terms of its components with respect to the matrix coefficients of the group representations.

Theorem H.2 *The irreducible representation functions $D^j(g)^m{}_n$ form a complete basis of (Lebesgue) square-integrable functions defined on the group manifold.*

© Springer Nature Switzerland AG 2024
S. Bilson-Thompson, *Loop Quantum Gravity for the Bewildered*,
https://doi.org/10.1007/978-3-031-43452-5

Any such function $f(g)$ can then be expanded as

$$f(g) = \sum_{j;mn} f_j{}^{mn} D^j(g)_{mn} \tag{H.2}$$

where $f_j{}^{mn}$ are constants which can be determined by inserting the above expression for $f(g)$ in Eq. (H.1) and integrating over the group manifold. Thus we obtain

$$\int d\mu(g) f(g) D_j^\dagger(g)^{mn} = \sum_{j';m'n'} \int d\mu(g) f_{j'}{}^{m'n'} D^{j'}(g)_{m'n'} D_j^\dagger(g)^{mn}$$

$$= \sum_{j';m'n'} f_{j'}{}^{m'n'} \frac{n_G}{n_j} \delta^{j'}{}_j \delta^{n'}{}_n \delta^{m'}{}_m, \tag{H.3}$$

which gives us

$$f_j{}^{mn} = \sqrt{\frac{n_j}{n_G}} \int d\mu(g) f(g) D_j^\dagger(g)^{mn}. \tag{H.4}$$

Regge Calculus

Regge showed in 1961 that one could obtain the continuum action of general relativity "in 2+1 dimensions" from a discrete version thereof given by decomposing the spacetime manifold into a collection of tetrahedral simplices [13,14], with curvature corresponding to an excess or shortage of angle traversed around each vertex. For instance, as discussed in Sect. 6.1 a plane 2D surface can be tiled with equilateral triangles, with six such triangles meeting at each vertex. If one attempted to fit in a seventh triangle around a given vertex, thereby effectively increasing the number of degrees in a full circle, the only way it could be accommodated would be by curving the resulting surface. Similarly if one attempted to omit a triangle, thereby reducing the number of degrees in a full circle, the only way to join the edges of adjacent triangles would be to curve the surface they formed (clearly the addition or omission of more triangles leads to more extreme curvature). Hence when many such tetrahedral simplices are joined together, curvature of the resulting discrete manifold is represented by positive or negative deficit angles.

The Regge action for the ith tetrahedron is

$$S_i = \sum_{a=1}^{6} l_{i,a} \theta_{i,a} . \tag{I.1}$$

Here the sum over a is the sum over the edges of the tetrahedron. The terms $l_{i,a}$ and $\theta_{i,a}$ are the length of the edge and the dihedral deficit angle, respectively, *around* the ath edge of the ith tetrahedron.

The Regge action for a manifold built up by gluing such simplices together is simply the sum of the above expression over all N simplices

$$S_{\text{Regge}} = \sum_{i=1}^{N} S_i .$$

© Springer Nature Switzerland AG 2024
S. Bilson-Thompson, *Loop Quantum Gravity for the Bewildered*,
https://doi.org/10.1007/978-3-031-43452-5

It was later shown by Ponzano and Regge [15] that in the $j_i \gg 1$ limit, the $6 - j$ symbol corresponds to the cosine of the Regge action [16]

$$\begin{Bmatrix} j_1 & j_2 & j_3 \\ j_4 & j_5 & j_6 \end{Bmatrix} \sim \frac{1}{12\pi V} \cos\left(\sum_i j_i \theta_i + \frac{\pi}{4}\right).$$

Fibre Bundles

J

Much of the mathematical discussion about quantum gravity (especially the formulation of BF theory) is presented in the language of fibre bundles. This language can seem fairly abstract to the uninitiated. However it is important, and provides several nice visualisations of the mappings associated to gauge transformations, connections, and the like, so it is useful provide an overview of the topic.

As a first step, consider two copies of the real number line, which we will refer to as X and Y. These are just one-dimensional vector spaces. We recognise that an ordinary function of one variable, $y = f(x)$ with $x \in X$ and $y \in Y$ is *not simply* a mapping $X \to Y$, but rather it is a mapping from X to a two-dimensional manifold consisting of points labelled by the pairs (x, y), hence $f : X \to X \times Y$. We recognise that $f(x)$ defines the locus of points $(x, y) \in X \times Y$ satisfying the relationship $y = f(x)$. It is a simple matter to define an inverse mapping $f^{-1} : X \times Y \to X$, for instance one which "throws away" the y value, $(x, y) \mapsto x$. This is simply the act of projecting a point in $X \times Y$ down to a point in X.

Generalising these concepts, we can imagine a function which assigns elements v of a vector space V to each point in a manifold M. For concreteness, imagine V is the space of vectors in two dimensions. We can represent this as a disk. Points in this disk denote vectors, with the direction and distance of the point from the centre of the disk indicating the direction and magnitude of the vector (Fig. J.1). There is one copy of V associated to each point $x \in M$. The space formed by "stacking together" the disks associated with each point in the manifold is $E = M \times V$, and is called the *total space*. It is easy to define a mapping which takes any point in E to a point in M, which is the generalisation of f^{-1}, above. We denote such mappings $\pi : E \to M$. The set of all points in E projected by π down to a given $x \in M$ is the vector space V_x which we can think of as living at or, in some sense, "above" x.

In general we would consider topological spaces rather than vector spaces specifically, and so it is often sensible to replace V by another symbol such as F or E_x, but for our purposes it is sufficient to stick to considering vector spaces, and so we will retain the notation we have established.

© Springer Nature Switzerland AG 2024
S. Bilson-Thompson, *Loop Quantum Gravity for the Bewildered*,
https://doi.org/10.1007/978-3-031-43452-5

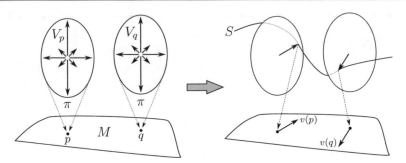

Fig. J.1 We associate the vector spaces V_p and V_q to points p and q in the manifold M. The mapping π projects all points in a given vector space down to the associated point in M (*left*). A section S through the "stack" of all vector spaces, $M \times V$, picks out a particular vector $v(x)$ in each V_x, and associates this to the corresponding point $x \in M$. A section therefore generalises the concept of a function, and defines a vector field on the manifold M (*right*)

Our choice to represent V as a disk is useful when talking about a two-dimensional vector space, but in general a more abstract representation is used. We will therefore adopt the usual practice of drawing E as a sheet above the manifold M, and representing V_x as a vertical strip running through this sheet (Fig. J.2). The collection of all such strips looks like a cluster of threads or fibres, so we call V_x a fibre. Just as a group is a set along with a binary operation (Appendix A), we refer to $E = M \times F$ along with a mapping $\pi : E \to M$ as a "fibre bundle". The total space is the union of all the fibres,

$$E = \bigcup_{x \in M} V_x .\tag{J.1}$$

In the cases under consideration, where the fibres are vector spaces, we call the fibre bundle a "vector bundle".[9] Assigning a vector $v(x) \in V_x$ to each point $x \in M$ means picking a point in E for each x. The locus of such points is S, a *section* of E. Choosing a section of E defines a vector field on M. We are then naturally drawn to consider the prospect of mappings between sections. Such mappings change one vector at any given point to another vector at the same point. We refer to these as *endomorphisms*. These are denoted End(V), and roughly speaking move us around from one place in V to another.[10] The collection of endomorphisms associated to all points in the manifold is called the "endomorphism bundle" End(E). A section T of End(E) defines a field of endomorphisms over M, which we can think of as linear transformations acting on the vector field defined on M. In other words, given a vector field on M (given by a section S of E), the section T of End(E) acts pointwise on it to produce a new vector field. If $T(p) \in G$, with G being a group that provides

---

[9] Actually, we also require the bundle to be *locally trivial*, but the precise meaning of this is a distraction from building a clear mental picture. See e.g. [2] for more details.

[10] Etymologically these are "inside/within-shape" mappings, because they stay within the vector space in question.

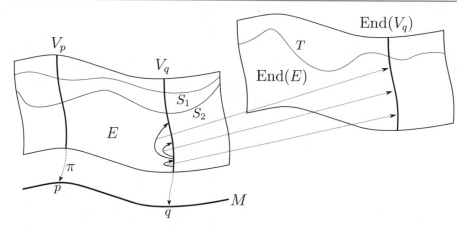

Fig. J.2 Sections of E determine vector fields on M according to where they intersect the fibres such as V_p and V_q. Endomorphisms, shown by the curved arrows adjacent to V_q, map between sections such as S_1 and S_2 and are themselves elements of a vector space, namely the fibre $\text{End}(V_q)$ of the bundle given by $\text{End}(E)$. A section T of $\text{End}(E)$ determines a change of vector field at each point p in M, and hence a gauge transformation

transformations of the vectors $v \in V$ then we can think of gauge transformations as sections of $\text{End}(E)$.

We should also note that A_μ, which we introduced in Eq. (3.4) in the context of the covariant derivative, can be viewed as an $\text{End}(E)$-valued 1-form. By this we mean to recognise that the role of A_μ is to tell us how a field (i.e. a section of E) defined on the manifold of interest changes when we move through the manifold in some direction. Thus when supplied with a vector field (defining the direction of motion) and a section S of E, the 1-form A_μ contracts with the vector field and enacts an endomorphism upon S to produce another field, S'.

Knots, Links, and the Kauffman Bracket

K

Our discussion in the main text, specifically Sect. 7.2.3, led us to consider closed curves which may be linked together, winding around each other (or if framed, themselves) several times. If we think of such closed curves as embedded in a manifold we can consider the prospect of distorting them and moving them around within the manifold. However, in a background-independent theory, the concept of distorting a curve loses its meaning—there is no metric against which to measure the positions of different parts of the curve. We are therefore drawn to consider topological features of such closed curves, or collections of closed curves, which would remain invariant if the curves were moved or deformed.

This leads us into a fairly wide area of knot theory, of which we will only scratch the surface. The reader may consult Chap. II.5 of [2], for instance, or specific texts on knot theory for more details.

In the main text we referred to "closed curves", as this terminology matches the language we used in discussing Wilson loops and spin networks. For now though, we will revert to the usual terminology of calling a closed curve a "knot". A knot is simply a 1-dimensional submanifold of $\mathbb{R}^3$ defined by a locus of points which can be mapped to the circle S^1. In other words, if you started at some point in the knot and moved from one point to another you would eventually return to your starting point without having intersected any of the points you had previously visited along the way. The simplest example is the *unknot*, which when projected into $\mathbb{R}^2$ does not cross itself at any point (Fig. K.1a). More complicated possibilities may occur, such as the trefoil knot, Fig. K.1b, which crosses itself three times when projected into $\mathbb{R}^2$. In any case, we can think of a knot as an embedding $S^1 \to \mathbb{R}^3$. A knot may be unoriented, or may have an orientation (i.e. a preferred direction around the knot), indicated by arrowheads drawn facing tangentially along the knot.

Given multiple knots we can combine them to form links. Several knots which do not cross at any point when projected into $\mathbb{R}^2$ constitute a perfectly valid (if somewhat trivial) link, but in general links involve sets of knots which cannot be projected into a plane without some crossings. The Hopf link (the union of two unknots) is shown

© Springer Nature Switzerland AG 2024
S. Bilson-Thompson, *Loop Quantum Gravity for the Bewildered*,
https://doi.org/10.1007/978-3-031-43452-5

(a) The unknot, without orientation (*left*), and with orientation (*right*)

(b) The trefoil knot, with orientation

Fig. K.1 The unknot and trefoil knot. The diagram of the unknot projected into two dimensions has no crossings, but for other knots any crossings are indicated with breaks to show which segment of the knot passes "under" or "behind" the other

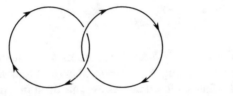 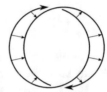

(a) The Hopf link, with orientation

(b) An oriented framed unknot, with a 2π twist

Fig. K.2 The linking of knots, and the self-linking (or "writhe") of a framed knot are closely related. In this case we can see that the two edges of the framed unknot **b** are linked through each other in exactly the same way as the component knots of the Hopf link **a**

in Fig. K.2a. A similar structure can be formed by framing the unknot. As mentioned in Sect. 7.2.3, framing a knot means we assign a non-tangential vector to every point of the knot, and let the tips of these vectors be the points defining a new knot. We then think of the knot as being stretched out into a ribbon with the original and newly-defined knots constituting the "left" and "right" edges of this ribbon. It is entirely possible that the framing involves twists through multiples of 2π (Fig. K.2b).

We can formalise the discussion above about distorting knots by considering the Reidemeister moves. These are a set of three ways in which crossings in a diagram can be changed, without cutting or rejoining the associated knot (strictly speaking there is a zeroeth Reidemeister move, but it does not involve crossings). Any two knot diagrams which can be transformed into each other via a sequence of Reidemeister moves correspond to knots which can be deformed into each other without breaking, rejoining, or passing one segment of a knot through another in $\mathbb{R}^3$. We say such knots are isotopic. The Reidemeister moves are illustrated in Fig. K.3.

We are now in a position to start discussing some invariants of knots and links. The simplest place to start is with the linking number, $\mathcal{L}$ (not to be confused with a lagrangian density, despite the choice of symbol!), equivalent to $\Phi[\lambda, \lambda']$ discussed in Sect. 7.2.3. As the name suggests, this just counts the number of linkings, and since two crossings must occur whenever a pair a knots K_1 and K_2 are linked, it is given by half the number of crossings. To make this more precise, consider a crossing of two

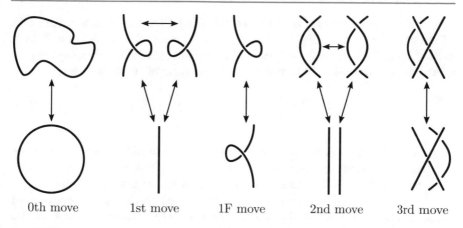

| 0th move | 1st move | 1F move | 2nd move | 3rd move |

Fig. K.3 The Reidemeister moves, including the modified first move (1F) which applies to framed knots

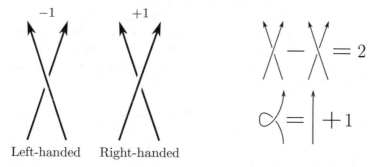

Left-handed   Right-handed

Fig. K.4 Left-handed and right-handed crossings of segments of oriented knots. If one imagines orienting one's hands so the thumbs point upwards, when knots form a left-handed (right-handed) crossing they wind around a vertical axis in the direction of the fingers of the left (right) hand. The sign of each crossing allows skein relations to be written for the writhe, where a diagram of a crossing stands in for the sign of that crossing (*right*)

oriented knots, and distinguish between left-handed and right-handed crossings, as per Fig. K.4. The linking number for a given link is half the number of right-handed crossings minus half the number or left-handed crossings (or in other words, let the *sign* of a left-handed crossing be -1 and the sign of a right-handed crossing be +1. The linking number is half the sum of these values for the entire link). You can confirm for yourself that the Hopf link as drawn in Fig. K.2a has linking number −1, though you may need to reorient the diagram to more easily determine if a given crossing is left-handed or right-handed. Confirm also that if the orientation of either of the component knots is reversed the linking number becomes +1, and reverts to −1 if the orientations of both knots are reversed.

An associated quantity which can be assigned to oriented framed knots is the self-linking number or writhe, $w(K)$. This is found by treating a framed knot as the link composed of its "left" and "right" edges, and counting the right-handed crossings minus the number of left-handed crossings. Thus, for example, the writhe of the framed unknot in Fig. K.2b is -2. A writhe can also be assigned to links as well as knots. In this case the writhe $w(L)$ is the sum of the linking numbers of pairs of knots in the link, plus the sum of the self-linking numbers of the knots,

$$w(L) = \sum_{i \neq j} \mathcal{L}(K_i, K_j) + \sum_i w(K_i) \tag{K.1}$$

We can consider how the writhe and linking number would change if we swapped left-handed crossings for right-handed crossings and vice-versa. This opens up the subject of *skein relations*. These are rules which tell us how a quantity like the linking number or writhe (or other more sophisticated invariants) change when we make modifications to a link diagram. In the simple case of the writhe, these take the form of algebraic relationships where the diagram of a crossing behaves equivalently to its numerical value. The skein relations can be used to find the writhe of a framed knot by successively undoing crossings (i.e. swapping a left-handed crossing for a right-handed crossing and vice-versa, and retaining an additive numerical term as per Fig. K.4), and using the Reidermeister moves (which leave the writhe unchanged) to convert the knot into an unknot (which has a writhe of zero).

Take note that if two links (or two knots) have different linking numbers or writhes they are not isotopic, but if they have the same linking numbers or writhes that does not necessarily mean they are isotopic.

A vast amount can (and has) been said about knots and links which will not be repeated here. We will simply mention another characterisation of links which is relevant to the discussion in Sect. 7.2.3, the Kauffman bracket $\langle L \rangle$ of a link L. This provides an algebraic expression which is determined via a state sum much like we would encounter in classical statistical mechanics. In this case, the sum is taken over possible values (named A and B, not to be confused with the vector potential and magnetic field!) assigned to the crossings of a link, and dependent on a parameter d which is raised to the power of the number of separate closed loops the link would form if all crossings were removed (this of course depends on exactly *how* each crossing is removed). With the specific choices $B = A^{-1}$ and $d = -(A^2 + A^{-2})$ the Kauffman bracket is found to be invariant under the permissible deformations of a framed link (i.e. the Reidemeister moves, with the first move being replaced by its framed equivalent).

The Kauffman bracket of a link is found by examining each crossing, facing such that it resembles a right-handed crossing (Fig. K.4), but without considering orientation. One then looks at all the ways the values A and B could be assigned to these crossings (i.e. for N crossings there are 2^N possible assignments). If a crossing is assigned A, it is replaced by a pair of uncrossed vertical lines. If a crossing is assigned B, it is replaced by a pair of uncrossed horizontal lines. Each assignment of As and Bs then produces a diagram consisting of d disjoint unknots. The polynomial assigned to $\langle L \rangle$ is found by adding the products of As and Bs for each assignment, weighted by the relevant power of d in each case. This is illustrated in Fig. K.5. The Kauffman bracket has its own particular skein relations which are also illustrated.

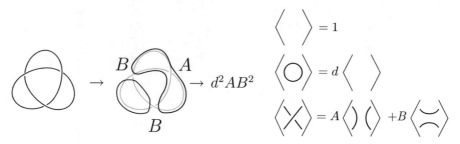

Fig. K.5 Calculation of one term (of eight) in the sum for the Kauffman bracket for the unoriented trefoil knot (*left*). The skein relations for the Kauffman bracket (*right*)

Quantum (or q-Deformed) Groups

Probably the first important thing to say about quantum groups is that they should really have a different name. It is not immediately obvious how (if at all) one would relate them to quantum mechanics, and, moreover they are noncommutative algebras, rather than groups. The term "q-deformed" is sometimes used, and gives a better indication of the means by which these structures are established.

To keep the discussion brief we will follow [17] to present a simple example, which illustrates the role of the deformation parameter q. Consider 2×2 matrices M and ϵ such that

$$
M = \begin{pmatrix} \alpha & \beta \\ \gamma & \delta \end{pmatrix}, \quad \epsilon = \begin{pmatrix} 0 & q^{-1/2} \\ -q^{1/2} & 0 \end{pmatrix}. \tag{L.1}
$$

We recognise that $\epsilon^2 = -\mathbb{1}$. We will not assume that α, β, γ, δ commute under multiplication. If we impose the condition

$$
M \epsilon M^\mathsf{T} = M^\mathsf{T} \epsilon M = \epsilon \tag{L.2}
$$

it is a simple matter to run through expanding the matrices and find that,

$$
\begin{aligned}
\alpha\beta &= q\beta\alpha & \beta\delta &= q\delta\beta & \delta\alpha - q^{-1}\gamma\beta &= 1 \\
\gamma\delta &= q\delta\gamma & \alpha\gamma &= q\gamma\alpha & \alpha\delta - q\gamma\beta &= 1 \\
& & \gamma\beta &= \beta\gamma
\end{aligned} \tag{L.3}
$$

These are clearly different to the relationships we would expect if the components of M were simply complex numbers and hence commuted. For example, we can easily show that

$$
[\alpha, \delta] = (q - q^{-1})\gamma\beta. \tag{L.4}
$$

© Springer Nature Switzerland AG 2024
S. Bilson-Thompson, *Loop Quantum Gravity for the Bewildered*,
https://doi.org/10.1007/978-3-031-43452-5

Notice that $\alpha\delta - q\gamma\beta = 1$ looks like the determinant of M, if it were in SL(2), except for that factor of q. We see that these relationships amount to a "deformation" of the relationships we would expect, which can be recovered by setting $q \to 1$. In this context the parameter q defines how much the quantum group is deformed from being what we would think of as an ordinary group (hence we can take a group like U(1) and deform it into a related quantum group) with $q = 1$ resulting in no deformation. Since doing so leads to $\alpha\beta = \beta\alpha$ etc., we see that Eqs. (L.3) are a result of choosing $q \neq 1$, and the supposition that the α, β, γ, δ do not commute under multiplication is secondary to the introduction of a deformation parameter. Note also that in the discussion of BF theories with a cosmological constant in Sect. 7.2.3 we saw that q was given by the exponential of the inverse of the Chern-Simons level, $q \sim \exp(1/k)$. Since k is inversely-proportional to Λ in such theories, we find that $q \to 1$ as $\Lambda \to 0$.

We can further explore the idea of a deformation (and indeed, something of the reason behind the name "quantum groups") by considering the transition from classical mechanics to quantum mechanics. Classical observables of a system, such as position and momentum, define points in phase space, and we can define an algebra of classical observables by reference to the Poisson bracket which serves as a binary operation (Sect. A.1). The transition to quantum mechanics is achieved by reference to operators corresponding to observables, and promoting the Poisson brackets to commutators, as per the discussion in Sect. 4.3, to obtain for example the Heisenberg uncertainty relation $[\hat{x}, \hat{p}] = \mathbf{i}\hbar$. We can consider this a deformation of the commutative algebra of classical observables into a noncommutative algebra, with a deformation parameter $\hbar$. When $\hbar \to 0$ we recover the classical world from the quantum. In a similar vein, we may use Eqs. (L.3) to write relations such as

$$\alpha\beta - \beta\alpha = [\alpha, \beta] = (q - 1)\beta\alpha, \tag{L.5}$$

that is to say the commutator $[\alpha, \beta] \neq 0$ when $q \neq 1$. This is the source of the name "quantum group", as the commutation relationship is reminiscent of the commutation relations between position and momentum in the Heisenberg uncertainty relation, above.

While q-deformed groups are too vast a subject to discuss in great detail here, it is worth becoming familiar with the foundational concepts and some terminology. The next few paragraphs provide an overview, drawn specifically from the discussions in [18–21], to which the reader is referred for more details.

The phase space of a system is usually thought of as a kind of abstract manifold representing the possible states the system can be in. But it can be endowed with extra structure, such that it becomes a group. That is to say, if g and h are points in the phase space G, we can define a mapping $G \times G \to G$ which obeys the closure, identity, inverse, and associativity properties (Sect. A.1) which turns the set G into a group.

We can treat the functions f on elements of G (which are mappings from elements of G to some field k, such as the complex numbers, for instance) as elements of an algebra, A. Then just as a group has an identity and inverse we find that there exists a mapping $\varepsilon : A \to k$ defined in terms of the identity,

$$\varepsilon(f) = f(e) \tag{L.6}$$

referred to as the *counit*. That is, the counit is a mapping that acts upon the functions f to yield the k-valued element that f would map the identity to. Similarly there is a mapping S defined in terms of the inverse which maps $A \to A$, called the *antipode*, such that

$$S(f)(g) = f(g^{-1}), \tag{L.7}$$

which can be visualised as acting on f to transpose where in k any two points g and g^{-1} are mapped to by f. In fact, in general there is a *coproduct*, corresponding to the product in G, denoted by Δ and mapping $A \to A \otimes A$ such that

$$\Delta(f)(g, h) = f(gh) = f_{(1)}(g) \otimes f_{(2)}(h) \tag{L.8}$$

(the subscripts referring to the two copies of A). If we have a mapping m on A, such that $m : A \otimes A \to A$, then Δ maps "back the other way". In short, given some group we can associate an algebra to that group and construct a coalgebra, equipped with the coproduct, counit, and antipode, corresponding to the product, unit, and inverse. Hence if

$$A \otimes A \otimes A \xrightarrow{m \otimes \mathbb{1}} A \otimes A \xrightarrow{m} A \tag{L.9}$$

we obtain a coalgebra C by reversing arrows, and replacing the product with the coproduct, hence

$$C \otimes C \otimes C \xleftarrow{\Delta \otimes \mathbb{1}} C \otimes C \xleftarrow{m} C \tag{L.10}$$

as in the figures on pages 6 and 11 of [19]. In this sense coalgebras are dual to algebras. Notice that in Eq. (L.9) we could have replaced $m \otimes \mathbb{1}$ with $\mathbb{1} \otimes m$, and in Eq. (L.10) we could have replaced $\Delta \otimes \mathbb{1}$ with $\mathbb{1} \otimes \Delta$, due to the associativity of m and Δ.

One can also show that ε and Δ are morphisms of algebras, and along with S they satisfy coinvertibility, such that

$$m \circ (S \otimes \mathbb{1}) \circ \Delta(f) = m \circ (\mathbb{1} \otimes S) \circ \Delta(f) = \varepsilon(f) \tag{L.11}$$

where $\circ$ denotes composition of mappings.

An algebra A which contains a unit element, and satisfies these conditions is referred to as a Hopf algebra. The name 'quantum group' is generally assigned to such algebras which are not commutative, nor cocommutative.

Entropy

M

Without thinking too hard about gases, and molecules moving at different speeds, entropy can simply be related to how much information is conveyed by learning that a system is in a given state. This may be seen as a measure of how surprising that state is. For instance, considering a fair coin toss, it takes one bit of information to say whether the coin lands on heads or tails. If the coin was two-headed, there would be no surprise associated with the fact that it landed on heads, and no information conveyed by saying the outcome of tossing this coin was heads. If the coin had a head and a tail, but was biased such that the probability of heads was greater than the probability of tails, there would be more surprise associated with reporting that it landed on tails than on heads, and more information conveyed by doing so. The fact that more information is conveyed by reporting the less likely outcome can be understood by analogy to asking the question "Is today a week day?" Assuming the day on which you ask the question is chosen randomly, there is a 2/7 chance the answer is no, and a 5/7 chance the answer is yes. The more likely answer doesn't convey much information, as it is not very specific and doesn't eliminate very many options (telling you that it's either Monday, Tuesday, Wednesday, Thursday, or Friday). With five remaining options you would need more than a single bit of information to specify exactly which day it is. But the less likely answer is more specific. You now need only a single bit of information to work out whether it's Saturday or Sunday, and hence the answer "No, today is not a weekday" carries more information than the answer "Yes, today is a weekday".

The question about whether today is a weekday is an example of a binary search. A range of options were split into two categories, and we asked a question which made us focus on one of those categories. In this case the categories were unequally sized, but in general a binary search would proceed in the following manner. Suppose we are trying to identify a target value x of some variable, which we know takes values in the set $\{x_i\}$ where $i = 1, \ldots, n$. For instance, the x_i may be the integers $i = 1, \ldots, n$, they may be the days of the week, they may be letters from the Latin alphabet, etc. We first ask whether i is greater than $n/2$. Suppose the answer is yes.

© Springer Nature Switzerland AG 2024
S. Bilson-Thompson, *Loop Quantum Gravity for the Bewildered*,
https://doi.org/10.1007/978-3-031-43452-5

We would then ignore the values less than $n/2$, and ask whether i is greater than $3n/4$ (i.e. which half of the range $n/2, \ldots, n$ does it lie in?). We would continue like this, dividing the range in half each time until we arrived at the value of x we were seeking. At each step we ask a simple binary question—"Is x in the top or bottom half of the remaining range?"—and hence to specify the location of x we need a number of bits equal to the number of questions we ask. How many questions do we need? We can deduce this by working backwards. The last question should decide between two options, that is 2^1 options. The penultimate question should subdivide at most four options, that is 2^2 options, in to equal halves. The third last question should divide 2^3 options into equal halves. And so on. If the total number of questions is Q, and x lies in the range $1, \ldots, n$ it follows that $n \approx 2^Q$ (remembering that in the real world, n won't necessarily be an integer power of 2, so we can expect approximate rather than exact equality). In other words, Q, which is the number of bits of information needed to specify x, is of order $\log_2 n$. The appearance of a logarithm now seems quite natural (pun intended, although in this case it of course not natural, but rather to the base 2).

Consider now the case of multiple fair coin tosses, for instance tossing two coins, each of which has a 50% probability of landing heads (H) or tails (T)—or equivalently tossing a single fair coin twice. The four possible outcomes are TT, TH, HT, and HH, each with equal probability of $1/4$. When the exact condition of each toss is specified, we call this a microstate. If we ignore any ordering of the coins we can then define three states, namely two-tails, one-head-one-tail, and two-heads, with respective probabilities $1/4$, $1/2$, and $1/4$. Since the exact outcome of each toss is not specified, we distinguish these from microstates, and call them macrostates. The one-head-one-tail macrostate is twice as probable as either of the other two macrostates because it contains two microstates (HT and TH), while the other macrostates contain only a single microstate each.[11]

In this case we see that it takes two bits of information to define each microstate. And hence it takes two bits of information to specify that the system is in either the two-tails or two-heads macrostates. But the one-of-each macrostate consists of two microstates, corresponding to a single distinguishing bit of information that we don't need. So only $2 - 1 = 1$ bit of information is needed to specify that the system is in this macrostate. Instead of asking two questions ("What was the result of the first toss?" and "What was the result of the second toss?") we could ask "Are both coins showing the same face?". Half the time the answer wil be yes, and no further questions are needed to find the macrostate of the system. The average information needed to specify which macrostate the system is in is the weighted mean of the number of bits for each macrostate, thus

$$I = \left(\frac{1}{2}(1 \text{ bit})\right)_{\text{one each}} + \left(\frac{1}{4}(2 \text{ bits})\right)_{\text{two heads}} + \left(\frac{1}{4}(2 \text{ bits})\right)_{\text{two tails}} = \frac{3}{2} \text{ bits}$$

$$\text{(M.1)}$$

---

[11] This is, incidentally, the basis for the traditional Australian gambling game "Two-up".

For a more complex example, consider the analogous situation when tossing four coins. This time four bits are needed to specify a particular microstate, as there are $2^4 = 16$ microstates, TTTT, HTTT, THTT, TTHT, TTTH, HHTT, HTHT,..., HHHT, HHHH in total. The macrostate corresponding to one head and three tails contains four microstates. Specifying a particular microstate within this macrostate requires two bits (which are redundant if we only want to specify the macrostate), and hence to answer the question "Is the system in the one-head-three-tails macrostate?" only requires $4 - 2 = 2$ bits of information. The number of redundant questions is the logarithm (to the base 2) of the number of microstates in the macrostate of interest.

Looking at the two-coin and four-coin examples, if we denote the number of microstates in our macrostate of interest by Ω_I, and the total number of microstates by $n = 2^Q$ as before, then the number of bits needed to specify a macrostate was equal to

$$Q - \log_2(\Omega_I) = \log_2(n) - \log_2(\Omega_I) = \log_2\left(\frac{n}{\Omega_I}\right) = \log_2(1/p_I) \qquad \text{(M.2)}$$

where p_I is the probability that the system will be in our macrostate of interest, if all microstates occur with equal probability. As above, the average information needed to specify which macrostate the system is in is the weighted mean of the number of bits for each macrostate, so we find

$$I = \sum_i p_i \log_2(1/p_i) = -\sum_i p_i \log_2(p_i). \qquad \text{(M.3)}$$

To keep things easy here, for pedagogical purposes, we have stuck to cases where each probability is a power of $1/2$, and used logarithms to the base 2. Of course, we might be tempted to look at examples where the probabilities were powers of $1/3$ and use logarithms to the base 3, or some other choice, in which case we're no longer counting bits, but some other unit of information (*trits* in the case of logarithms to the base 3, *nats* in the case of natural logarithms, etc.) but that would not fundamentally change our result, as we can always convert between logarithms to the base 2 and logarithms to another base using the well-known formula,

$$\log_a(x) = \frac{\log_b(x)}{\log_b(a)} \qquad \text{(M.4)}$$

and hence convert our answer to bits. This shows us that Eq. (M.3) is just Shannon's entropy formula, Eq. (8.4), if we identify the information needed to specify a macrotate with the entropy associated to that macrostate.

The connection to thermodynamic entropy can be seen as follows. Imagine that we have a container, divided into equal subsections labelled A, B, C, etc. The subsections are separated by impermeable partitions. Initially, there are N gas molecules contained in subsection A. At some time the partitions are removed, allowing the gas molecules to diffuse to other subsections. If we were to ask where gas molecule 1, gas molecule 2,..., gas molecule N are located at the moment the partitions are removed we will get the entirely straightforward answer AAAAAAAA... (a string of As, of length N). The container is, in essence, a machine that produces a completely predictable string of symbols from the list A, B, C,...

A short while later, some of the molecules will have diffused to the other subsections, but most will still remain in subsection A. If we were to ask the locations of the gas molecules now we might get a string like AAABAACA... The container has become a machine which produces a fairly predictable string of symbols. That is, given each symbol in the string, we have a good (but not perfect) chance of predicting the next symbol, since we know that it will most likely be an A.

Eventually the gas molecules will diffuse until they are equally (or very nearly so) distributed among the subsections of the container. Now we would obtain a string like EABBFCEDF... The container is now a kind of machine that produces an entirely random answer, in which the probability of each symbol is equal and independent of the symbols that came before it.

The crucial point is that each gas molecule is like a coin (or a coin toss) in our examples above. And the subsections we may find the molecules in are like the outcomes—heads or tails—of a coin toss. Each string of symbols corresponds to a microstate of the gas molecules in the container, just as a string of symbols like HTTH corresponded to a microstate of coin tosses. There is only one way to create a microstate in which all the molecules are in subsection A. But there are multiple ways (microstates) to have the molecules distributed between several subsections. The number of microstates corresponding to the molecules being equally distributed is larger than for any other situation. For instance, with three subsections A, B, C, and three molecules the possibilities are;

	0 Cs	1 C	2 Cs	3 Cs
0 Bs	AAA	CAA, ACA, AAC	CCA, CAC, ACC	CCC
1 B	BAA, ABA, AAB	ABC, ACB, BAC, BCA, CAB, CBA	CCB, CBC, BCC	
2 Bs	BBA, BAB, ABB	BBC, BCB, CBB		
3 Bs	BBB			

Hence if each microstate is equally likely we would expect the system to more often be in microstates corresponding to the molecules being distributed between multiple subsections, rather than all concentrated into one subsection, since there are more of the former microstates than the latter. The probability of any given macrostate is proportional to the number of microstates it contains. In this example, the macrostate consisting of all three molecules being in different subsections contains six microstates, and hence it is twice as likely as the macrostate with two molecules in A and one molecule in B, etc. As the number of molecules goes up, the probabilities of macrostates in which the molecules are fairly evenly distributed rapidly outstrips the probabilities of macrostates where they are unevenly distributed.

It should now be clear to see that the evolution from an uneven distribution of gas molecules to an even distribution may be thought of as the container of gas evolving from being a machine that outputs a low-entropy, predictable, unsurprising, string of characters, such as AAAAA... (we'll call such a machine a low-entropy macrostate) to being a machine that outputs a high-entropy, unpredictable, surprising string of characters (and we'll call such a machine a high-entropy macrostate). Furthermore, this evolution is a probabilistic process, determined by the fact that the system is more likely to be in a macrostate which contains many microstates rather than one which contains few microstates. As this process is probabilistic, we expect it to apply to numerous other physical systems besides containers of gas—including the complex, fluctuating networks of spins that describe spacetime in loop quantum gravity.

Square-Free Numbers

According to the fundamental theorem of arithmetic, any integer $d \in \mathbb{Z}$, has a unique factorization in term of prime numbers,

$$d = \prod_{i=1}^{N} p_i^{m_i}$$

where $\{p_1, p_2, \ldots, p_N\}$ are the N prime numbers which divide d, one or more times, and m_i is the number of times the prime number p_i occurs in the factorization of d. Thus we have

$$\sqrt{d} = \prod_{i=1}^{N} p_i^{m_i/2}.$$

We can partition the set $\{m_i\}$ into two sets containing only the even or odd elements respectively

$$\{m_i\} \equiv \{m_j^e\} \cup \{m_k^o\}$$

where $j \in 1 \ldots n_e$, $k \in 1 \ldots n_o$, and $n_e + n_o = N$. This gives

$$\sqrt{d} = \left(\prod_{i=1}^{n_e} p_i^{\frac{m_i^e}{2}} \right) \left(\prod_{j=1}^{n_o} p_j^{\frac{m_j^o}{2}} \right).$$

Since each of the $m_i^e = 2a_i^e$ and $m_j^o = 2b_j + 1$, for some $a_i, b_j \in \mathbb{Z}$, we have

$$\sqrt{d} = \left(\prod_{i=1}^{n_e} p_i^{a_i^e} \right) \left(\prod_{j=1}^{n_o} p_j^{m_j^o} \right) \sqrt{\prod_{k=1}^{n_o} p_k} = A\sqrt{B}.$$

© Springer Nature Switzerland AG 2024
S. Bilson-Thompson, *Loop Quantum Gravity for the Bewildered*,
https://doi.org/10.1007/978-3-031-43452-5

It is evident that since the third term in the product has no repeating elements, its square-root $\sqrt{B}$ cannot be an integer (i.e. the presence of repeating elements would lead to an expression like $\sqrt{X \cdot X}$). Such an integer B is therefore known as a square-free integer. Thus any integer d can be written as the product of a square-free integer B and another (non square-free) integer $C = A^2$ such that $d = C \times B$.

Brahmagupta-Pell Equation

O

Around the 7th century A.D. the Indian mathematician Brahmagupta demonstrated the Brahmagupta-Fibonacci Identity,

$$(a^2 + nb^2)(c^2 + nd^2) = (ac)^2 + n^2(bd)^2 + n[(ad)^2 + (bc)^2]$$

$$+2acbdn - 2acbdn \qquad (O.1)$$

$$= (ac + nbd)^2 + n(ad - bc)^2 \qquad (O.2)$$

where we have added and subtracted $2acbdn$ from the left-hand side on the first line. The above goes through for all $n \in \mathbb{Z}$. Given any pair of triples of the form (x_i, y_i, k_i), where $i = 1, 2$, which are solutions of the Diophantine equation $x_i^2 - ny_i^2 = k_i^2$, we can construct a third triple (x_3, y_3, k_3), which is also a solution of the same equation, by applying the Brahmagupta-Fibonacci identity to the first two pairs

$$(x_1^2 - ny_1^2)(x_2^2 - ny_2^2) = (x_1 x_2 - ny_1 y_2)^2 - n(x_1 y_2 - x_2 y_1)^2 \qquad (O.3)$$

which tells us that $x_3 = x_1 x_2 - ny_1 y_2$, $y_3 = x_1 y_2 - x_2 y_1$ and $k_3 = k_1 k_2$. One can easily check that the triple $\{x_3, y_3, k_3\}$ is also a solution of the Diophantine equation.

When we apply the restriction that $k_i = 1$, the Diophantine equation $x_i^2 - ny_i^2 = k_i^2$ reduces to the BP equation,

$$x_i^2 - ny_i^2 = 1$$

and given two pairs of solutions $\{(x_i, y_i), (x_j, y_j)\}$ to the BP equation (for the same fixed value of n), we can generate a third solution given by

$$(x_k, y_k) = ((x_i x_j - ny_i y_j), (x_i y_j - x_j y_i)). \qquad (O.4)$$

© Springer Nature Switzerland AG 2024
S. Bilson-Thompson, *Loop Quantum Gravity for the Bewildered*,
https://doi.org/10.1007/978-3-031-43452-5

More generally given any solution $(x_0, y_0; n)$ to the BP equation, one can generate an infinite set of solutions $(x_i, y_i; n)$ by repeatedly applying the Brahmagupta-Fibonacci identity to the starting solution

$$(x_1, y_1) = (x_0, y_0)^2$$
$$(x_2, y_2) = (x_0, y_0)(x_1, y_1)$$
$$\vdots$$
$$(x_n, y_n) = (x_0, y_0)(x_{n-1}, y_{n-1}). \tag{O.5}$$

Here, the pair $(x_0, y_0; n)$ is referred to as the fundamental solution.

O.1 Quadratic Integers and the BP Equation

We are familiar with solutions of equations of the form

$$x^2 + Bx + C = 0.$$

This is the quadratic equation from beginning algebra courses, which has as solutions

$$x_\pm = \frac{-B \pm \sqrt{B^2 - 4C}}{2}$$

when the *discriminant* $B^2 - 4C$ is negative, the roots of the equation are imaginary or complex numbers

$$x_\pm = \frac{-B \pm i\sqrt{D}}{2} \in \mathbb{C}$$

where $D = |B^2 - 4C|$ and $i = \sqrt{-1}$. When $\{B, C\} \in \mathbb{Z}$, the solutions of the quadratic equations can be characterized as elements of the field of quadratic integers $\mathbb{Q}(\sqrt{D})$, which is an extension of the familiar field of rational numbers $\mathbb{Q}$. Such numbers have the form

$$z = a + \omega b$$

where $\{a, b\} \in \mathbb{Z}$, $\omega = \sqrt{D}$ if $d \bmod 4 \equiv 2, 3$ and $\omega = \frac{1+\sqrt{D}}{2}$ otherwise (if $D \bmod 4 \equiv 1$). Note that $D \in \mathbb{A}$, where $\mathbb{A}$ is the set of square-free integers.

It is at this point that one makes a connection to the square-free quadratic extension of the field of rationals $\mathbb{Q}(\sqrt{n})$ and its integral subset $\mathbb{Z}(\sqrt{n})$, by noting that any solution $(x_i, y_i; n)$ of the BP equation can be represented as a quadratic integer,

$$(x_i, y_i; n) \Rightarrow z_i^n = x_i + y_i \sqrt{n} \in \mathbb{Z}(\sqrt{n}).$$

The consistency of this representation is enforced by the fact that the multiplication law for two quadratic integers $z_i, z_j \in \mathbb{Z}(\sqrt{n})$ is the same condition satisfied when multiplying two pairs of solutions of the BP equation to obtain a third pair, i.e. , if

$z_i = x_i + y_i \sqrt{n}$ and $z_j = x_j + y_j \sqrt{n}$ are two members of $\mathbb{Z}(\sqrt{n})$, then their product $z_k = z_i \times z_j = x_k + y_k \sqrt{n}$ is given by

$$x_k = x_i x_j + n y_i y_j \tag{O.6}$$

$$y_k = x_i y_j + x_j y_i \tag{O.7}$$

which is identical to the multiplication law satisfied by pairs of solutions of the BP equation.

References

1. M.E. Peskin, D.V. Schroeder, *An Introduction to Quantum Field Theory* (Addison Wesley, 1995). isbn: 9780201503975. https://doi.org/10.1201/9780429503559
2. J.C. Baez, J.P. Muniain, *Gauge Fields, Knots, and Gravity*, vol. 4. Series on Knots and Everything (World Scientific Pub Co Inc, Oct. 1994). isbn: 9810220340. https://doi.org/10.1142/2324
3. R.M. Wald, *General Relativity* (The University of Chicago Press, 1984). isbn: 9780226870335. https://doi.org/10.7208/chicago/9780226870373.001.0001
4. S.H.S. Alexander, D. Vaid, *Gravity Induced Chiral Condensate Formation and the Cosmological Constant* (Sept. 2006). https://doi.org/10.48550/arXiv.hep-th/0609066. arXiv:hep-th/0609066
5. S.M. Carroll, Lecture Notes on General Relativity (Dec. 1997). https://doi.org/10.48550/arXiv.gr-qc/9712019. arXiv:gr-qc/9712019
6. M. Bojowald, *Canonical Gravity and Applications: Cosmology, Black Holes, Quantum Gravity* (Cambridge University Press, 2010). https://doi.org/10.1017/CBO9780511921759
7. Joseph D. Romano, Geometrodynamics vs. Connection Dynamics. Gen. Rel. Grav. **25**, 759–854 (1993). https://doi.org/10.1007/BF00758384. (arXiv: gr-qc/9303032)
8. E. Witten, A note on the Chern-Simons and Kodama wavefunctions (June 2003). https://doi.org/10.48550/arXiv.gr-qc/0306083. arXiv:gr-qc/0306083
9. A. Randono, Generalizing the Kodama state I: construction (Nov. 2006). https://doi.org/10.48550/arXiv.gr-qc/0611073. arXiv:gr-qc/0611073
10. A. Randono, Generalizing the Kodama state II: properties and physical interpretation (Nov. 2006). https://doi.org/10.48550/arXiv.gr-qc/0611074. arXiv:gr-qc/0611074
11. A. Randono, In Search of Quantum de Sitter Space: Generalizing the Kodama State. Ph.D. thesis. University of Texas at Austin, 2007. https://doi.org/10.48550/arXiv.0709.2905. http://arXiv.org/abs/0709.2905
12. W.-K. Tung, *Group Theory in Physics* (World Scientific Publishing Company, Sept. 1985). isbn: 9971966573. https://doi.org/10.1142/0097
13. T. Regge, General relativity without coordinates. Il Nuovo Cimento (1955–1965) **19**(3), 558–571 (1961). issn: 0029-6341. https://doi.org/10.1007/BF02733251. http://dx.doi.org/10.1007/BF02733251
14. J. Iwasaki, A reformulation of the Ponzano-Regge quantum gravity model in terms of surfaces (Oct. 1994). https://doi.org/10.48550/arXiv.gr-qc/9410010. arXiv:gr-qc/9410010
15. G. Ponzano, T. Regge, Semiclassical limit of Racah coefficients, in *Spectroscopic and Group Theoretical Methods in Physics*. ed. by F. Bloch et al. (Amsterdam, North-Holland, 1968), p. 1
16. T. Regge, R.M. Williams, Discrete structures in gravity. J. Math. Phys. **41**, 3964–3984 (2000). https://doi.org/10.1063/1.533333. arXiv:gr-qc/0012035
17. R.J. Finkelstein, Quantum groups and field theory. Mod. Phys. Lett. A **15**, 1706–1709 (2000). https://doi.org/10.1142/S0217732300002218. (arXiv: hep-th/0003189)
18. P. Etingof, M. Semenyakin, A brief introduction to quantum groups (2021). https://doi.org/10.48550/arXiv.2106.05252. arXiv:2106.05252v1

19. D. Manchon, Hopf algebras, from basics to applications to renormalization. Comptes Rendus des Rencontres Mathematiques de Glanon (2003). https://doi.org/10.48550/arXiv.math/0408405. arXiv:0408405
20. P. Clare, *Introduction to Quantum Groups I. Groups, Algebras and Duality* (2020). https://prclare.people.wm.edu/GAG/GAG_200930_Clare.pdf
21. E. Shelburne, *Introduction to Quantum Groups II. q-Deformations of Lie Algebras* (2020). https://prclare.people.wm.edu/GAG/GAG_201007_Shelburne.pdf

Printed in the United States
by Baker & Taylor Publisher Services